MAGIC SQUARES OF ORDER 3

S.VENKATACHALAM

MAGIC SQUARES OF ORDER 3

S.VENKATACHALAM
Consulting Engineer,
(Former Professor &Head, Dept. Of Met.Engg &Mat.Sci.
IIT,Bombay)

To

My grandchildren

Divya, Adit, and Swara

Preface

Magic squares form a very important part of 'Recreational Mathematics'. Magic squares have been known since ancient times in India and China. Later, the knowledge of magic squares was carried by Arabs to the West.

A magic square of order n is a square matrix consisting of numbers such that the sum of the elements of each row, column and diagonal is the same. This is called the 'magic sum'. Magic squares can be of any order. They are classed into odd order, singly even and doubly even order squares. The method of construction of odd order magic squares is simpler than the even order squares.

In this book we shall deal with only 3X3 magic squares. This is the simplest and first odd order magic square. Semi and hetero magic squares are also discussed. Calendar magic squares, navagrahas, prime numbers, alphamagic squares and palindromic magic squares find a place in the book. The contributions of Ramanujan and Robertson to magic squares are discussed. Apart from additive magic squares, multiplicative magic squares also form an important part of magic squares. Here, the products of the different elements of each row, column and diagonal are considered instead of addition. Instead of integers, fractions and decimals can also be used in magic squares. At the end of the book a quiz is given dealing with English vocabulary and Mathematics. Questions on magic squares form the concluding chapter.

I thank Ms. Manasi Lele and Ms.Harshada Kanade for painstakingly typing the manuscript. The C code was written by Ms. Harini Arun and the Python code was written by Dr. Manoj Kumar Singh. I am thankful to them.

Lastly I thank my wife Parvatham and my daughters Sugandha and Sangeeta for their support and encouragement throughout my journey into the world of magic squares.

Students and others with an interest in mathematics for entertainment will find this book very useful. Today magic squares are taught from the school level and researchers doing doctoral and post doctoral work in this field is not uncommon. If this book generates an interest in the reader and he starts exploring other areas of recreational mathematics, the purpose of writing this book would have been served.

S.Venkatachalam

CONTENTS

    C code and Python code for generating odd order magic squares

Chapter 1

INTRODUCTION

From ancient times, numbers have been considered to be endowed with magical powers. Some numbers were considered to have special properties. 7 was considered as a lucky number. Number 13 was considered as an unlucky number.

Magic square is an example of "magic" in numbers. According to legend, magic squares first appeared in China. We do not know whether this story is true. Legend tells us that in the 26th century before Christ, Chinese people brought sacrifices to their river God because the river almost flooded. The emperor Yu observed a divine turtle crawling out of the Yellow river. On the back of the turtle was a curious pattern of dots.

On close scrutiny, it was observed that the sum of the numbers in each row, column and diagonal was the same and it was 15. This square became known as the famous *lo-shu*. This is the normal magic square of order 3 in which 1 is at the bottom and 2 is in the upper right hand corner.

4	9	2
3	5	7
8	1	6

Mystical significance was attributed to this magic square. Even numbers were in the corners. They were thought to "symbolize" "the female-passive" or *ying*. The odd numbers were considered as "male-active" or *yang*. The number 5 in the centre represents earth surrounded by 4 major "elements" of metal (4 and 9), fire (2 and 7), water (1 and 6) and wood (3 and 8). All the elements contained both *ying* and *yang*, male and female.

Magic squares have been known in India since ancient times. Even to this day, you can find in Indian Tamil almanacs a Sita chakram and a Rama chakram. Sita chakram is nothing but the 3X3 magic square. Rama chakram is a 4X4 magic square.

2	9	4
7	5	3
6	1	8

9	16	5	4
7	2	11	14
12	13	8	1
6	3	10	15

Sri Sita chakram   Sri Rrama chakram

In the above squares, each number is associated with a prediction. The believer takes a small flower, prays to his deity and closing his eyes drops the flower on the chakra.

The number on which the flower falls is believed to give the prediction.

In Parshvanath Hindu-Jain temple (10th century) in Khajuraho in the state of Madhya Pradesh in India the *chautisa yantra* can be seen. *Chautisa* in Hindi means 34 which is the magic sum of this 4X4 magic square.

7	12	1	14
2	13	8	11
16	3	10	5
9	6	15	4

There are many ways in which you can add up four numbers and get the magic sum of 34. It will be interesting if you can list out all combinations which will yield the magic sum.

How to construct a 4X4 magic square has been very lucidly explained by Bhaskaracharya in the following verse in Sanskrit.

वाञ्छा कृतार्ध कृतमेकहीनम् ।
द्विके ग्रहे षोडशे सप्तमेष्टमे ॥
तिथ्यावतारे प्रथमेवशिष्टे ।
द्विसप्त षट त्रि अष्ट भू वेद प्राणाः ॥

You ask for the magic sum which is even and greater than 34. You may then proceed to construct the 4X4 magic square.

- Make a 4X4 grid and number them from 1 to 16 as shown
- Divide the magic sum by 2 and reduce 1 from the result.

- This number goes into cell number 2
- Continue reducing 1 from the previous result and fill the numbers in the following squares: 9, 16, 7, 8, 15, 10
- By now, 8 numbers would have been filled
- Fill the rest of the numbers in the following order: 2, 7, 6, 3, 8, 1, 4, 5

The following square has been constructed taking 40 as the magic sum of the square.

1	2	3	4
12	19	2	7
5 6	**6** 3	**7** 16	**8** 15
9 18	**10** 13	**11** 8	**12** 1
13 4	**14** 5	**15** 14	**16** 17

There are 40 ways in which can get the magic sum of 40. Try to find out yourself.

A 4X4 magic square connected with Ramanujan's life is given below.

22	12	18	87
66	50	19	4
41	23	47	28
10	54	55	20

The magic sum of this square is 139. Did you notice the peculiarity of this 4X4 magic square? Look at the first row. It gives the date of birth of Ramanujan which is 22nd December 1887. Magic squares like this can be constructed for other special occasions also like marriage date etc.

Magic squares were introduced to the west by Arabs after acquiring the knowledge from India. During the renaissance period, the mathematician Cornelius Agrippa (1486-1535) constructed magic squares of orders 3 to 9 to represent the then known planets.

Another famous example of a magic square which appeared in Albrecht Durer's engraving is Melancholia I. On a wall behind an angel is the following square.

Albrecht Durer

16	3	2	13
5	10	11	8
9	6	7	12
4	**15**	**14**	1

In 1514, Durer made this famous engraving. You may note that the date of the engraving 1514 appears in the last row.

Durer's magic square and his Melancholia I both played large roles in Dan Brown's 2009 novel, "The Lost Symbol". This 4X4 magic square also has the magic sum of 34.

Magic squares were seen as having marvelous powers. They were carved on to amulets and silver tablets for decoration and protection against the plague in the 16th and 17th centuries.

A magic square of order n is a square matrix or array of n^2 numbers such that the sum of the elements of each row, column as well as the main diagonal and the main back diagonal is the same number called the magic sum, magic constant or line sum. The main diagonal consists of entries from "north west to south east". By main back diagonal is meant the entries on the diagonal from "north east to south west".

Generally the entries are thought of as the natural numbers 1, 2, 3,.......n^2, where each number is used exactly once. Such squares are referred to as normal magic squares, classical magic squares, pure magic squares and traditional magic squares.

In addition to the main and back diagonal, the broken diagonals are also considered for the magic sum in magic squares. See the following 2 figures. The dots indicate the entries in the broken diagonals.

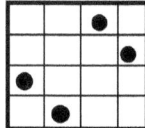   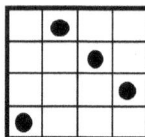

Such a magic square where the broken diagonals also sum to the magic constant is known as panmagic square, pandiagonal square or diabolic square.

Claude Gasperd de Meziriac (1518-1618) was the first to propose line paths that linked the numbers in numerical order.

We have so far seen some examples of magic squares of order 3 and 4. Magic squares can be of any higher order. Natural numbers can be classified as follows.

1. odd- not divisible by 2 like 3, 5, 7, 9,........

2. singly even- even and divisible by 2 and not 4 like 6, 10, 14.....

3. doubly even- this term is not very descriptive and all it means is a multiple of 4 like 4, 8, 12 and so on.

Magic squares can be constructed for all these orders. It is easy to construct odd numbered magic squares. To construct even order magic squares is a bit more challenging. Odd number magic squares (3X3, 5X5 and so on) can be generated by a very simple algorithm. It is the simplest and easiest to implement.

As order n increases, the number of magic squares increases rapidly. As far back as 1693, the 880 different 4th order magic squares were published posthumously by French mathematician Bernard Frenicle de Bessey (1602-1675).

Magic squares of different orders-how many solutions are there?

The 3X3 magic square has only one unique solution if you do not count the rotations and reflections. The following table gives you the number of solution upto 10X10 magic squares.

Order of magic square	Number of solutions
3X3	1
4X4	880
5X5	275,305,224
6X6	1.77×10^{19}
7X7	3.79×10^{34}
8X8	5.22×10^{54}
9X9	7.84×10^{79}
10X10	2.41×10^{110}

Special kinds of magic squares

Depending on the additional properties they possess, certain kinds of magic squares have been given more narrow definitions.

Semi magic squares

It is an n x n matrix, such that the sum of the elements on each row and column is equal. Nothing is required of the diagonals. One or both the diagonals may not sum to the magic constant. An example of semi magic squares of order 5 is shown below.

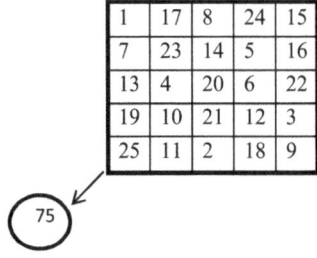

1	17	8	24	15
7	23	14	5	16
13	4	20	6	22
19	10	21	12	3
25	11	2	18	9

75

The rows, columns and main diagonal sums to the magic constant 65. Only the main back diagonal sums to 75.

Symmetric magic squares

In addition to being magic, it has the property that "the sum of the numbers in any 2 cells symmetrically placed with respect to the centre cell is the same. It is called as associative magic square. Multiplying n/2 by the sum of a pair of numbers symmetrically placed + centre square give the magic sum. Any magic square produced by Hindu (stair step) method will be symmetric.

A 5th order symmetrical and pandiagonal magic square is given below.

1	15	22	8	19
23	9	16	5	12
20	2	13	24	6
14	21	10	17	3
7	18	4	11	25

Hetero square

It is an arrangement of integers 1 to n^2 in a square such that the rows, columns and diagonals all sum to different values. There are no hetero squares of order 2, but hetero squares exist for any order $n \geq 3$. Examples are given for orders 3, 4 and 5.

1	2	3
8	9	4
7	6	5

2	1	3	4
5	6	7	8
9	10	11	12
13	14	15	16

1	2	3	4	5
16	17	18	19	6
15	24	25	20	7
14	23	22	21	8
13	12	11	10	9

Anti magic square

It is a special case of hetero square. Here, each row, column and main diagonal produce different sums such that these sums form sequence of consecutive integers. An example is given below.

21	18	6	17	4	→ 66
7	3	13	16	24	→ 63
5	20	23	11	1	→ 60
15	8	19	2	25	→ 69
14	12	9	22	10	→ 67

65   62   61   70   68   64   59

Concentric or bordered magic square

It is a magic square for which removing the top and bottom rows and the left and right columns (the borders) results in another magic square.

In the following example, a 5X5 magic square has a 3X3 square within it. When you remove the borders of the 5X5 magic square, you are left with a 3X3 magic square.

18	20	21	04	02
01	12	11	16	25
03	17	13	09	23
19	10	15	14	07
24	06	05	22	08

Zero magic square

It is a magic square whose magic constant is zero. It cannot be a normal magic square since it must contain negative entries. An example is given below.

4	11	-12	-5	2
10	-8	-6	1	3
-9	-7	0	7	9
-3	-1	6	8	-10
-2	5	12	-11	-4

Geometric or multiplicative magic square

So far we have seen additive magic squares where we looked at the magic sum. A multiplicative magic square is a square matrix of numbers such that the product of the elements of each row, column and diagonal is a constant. An example of a 4X4 square is given below.

432	6	18	16
4	72	24	108
8	36	12	216
54	48	144	2

Its magic product is 746,496.

Additive-multiplicative magic square

It is a magic square in which both the sum and product in each row, column and main diagonal and back diagonal is a constant. An 8X8 magic square is given below as an example.

162	207	51	26	133	120	116	25
105	152	100	29	138	243	39	34
92	27	91	136	45	38	150	261
57	30	174	225	108	23	119	104
58	75	171	90	17	52	216	161
13	68	184	189	50	87	135	114
200	203	15	76	117	102	46	81
153	78	54	69	232	175	19	60

The magic sum of this square is 840. The magic product is 2,058,068,231,856,000.

Latin squares

Latin squares are nXn arrays of the n elements such that the same element appears exactly once in any given row or column. There are certain types of Latin squares of interest in dealing with magic squares. A Latin square is diagonal provided each element appears exactly once in the main diagonal and back diagonal. Two Latin squares are said to be *orthogonal* provided that "if superimposed, every cell value of one square matches once and once only with every cell value of the other square".

Examples of 2 orthogonal diagonal Latin squares of order 4 are given below.

4	2	3	1
3	1	4	2
1	3	2	4
2	4	1	3

and

4	2	3	1
1	3	2	4
2	4	1	3
3	1	4	2

Methods of constructing magic squares

Several methods of constructing of magic squares exist. Squares of odd order have different construction methods from squares of even order. Even order squares may have different methods depending on whether or not the order is a multiple of 4 (called doubly even) or not (called singly even).

Odd magic squares

1. The Hindu method.

This is also known as the staircase method. Some say it was developed by De la Loubere. It need not use integers 1 through n^2. Any standard arithmetic sequence can generate a magic square.

2. Method of Bachet de Meziriac

It follows the same pattern as above but after each diagonal of n numbers, rather than moving one space down, the next number is placed 2 spaces to the right. This pattern does not begin in the centre position of top row.

3. Method of Phillipe de la Hire

To construct an odd ordered magic square of order n, first construct 2 n^{th} order Latin squares. The first Latin square consists of the numbers 1 through n in each row or column. The elements of the rows and columns of the second Latin square are 0 and the first (n-1) multiple of n. The 2 Latin squares must be orthogonal. The sum of the 2 Latin squares is a magic square. Magic squares produced by this method are pan diagonal as well.

4. Knight movement method

This very interesting system is somewhat similar to the staircase method and also requires the concept of imaginary squares. It involves the knight's moves as in chess. To those who are not familiar with chess, there are 8 moves a knight can make. Magic squares may be constructed using all the eight knight's moves depending on the placement of number 1.

The construction methods enumerated above will not be discussed here. All these methods are discussed in detail with several examples in the chapters that follow. The reader is advised to construct similar magic squares to get familiar with the different methods.

In the chapters that follow, we shall discuss magic squares of order 3 only. The special properties of these squares are discussed. All aspects of 3rd order squares involving prime numbers, composite numbers, palindromes and many more are explained. Various quizzes and questions are also given at the end of this book. The reader is encouraged to solve these.

Chapter 2

3X3 MAGIC SQUARES

For a person who is not familiar with magic squares, you may get him interested in this subject very easily. You draw a 3X3 grid and ask him to fill the nine boxes with numbers from 1 to 9. You also tell him that each number must be used only once. The result should be magic square which means that the numbers in 3 rows, 3 columns and 2 diagonals add up to the same total.

Your friend may think that this is a very simple problem and what should be the difficulty in filling out all the nine numbers. Once he attempts to do it, he will realize that this is not so simple. If you ask him what should be the magic sum of such a magic square using numbers from 1 to 9, he may find it difficult to answer that question.

If anyone thinks he can solve this problem by trial and error, he is mistaken. This can be explained as follows. How many ways are there to fill such a square? Let us think of the square as filled with one number at a time, starting with placing number 1 somewhere and ending with placing 9. There are 9 ways to place 1, followed by 8 ways to place 2, and so on until the last number 9 is placed in the only unoccupied cell of the 3X3 grid. Hence, there are 9! (9 factorial) which is 9x8x7.... x1 = 362,880 ways to arrange.

Assuming that you take 1 minute for filling the square and checking whether it is fulfils the condition, you will take 362, 880 minutes to examine all the possibilities. This works out to 6048 hours. If one works continuously 24 hours of the day, it will take 252 days. You will be surprised to know that there are only 8 possible solutions. As you will read later, even these 8 will be correct if you include rotations and reflections. Actually, there is only one unique solution. More of this is given later in this chapter.

3X3 magic square is the first possible magic square. You may wonder why 2X2 magic square is not possible. Although even order squares are not the subject matter of this book, still it is worthwhile discussing the impossibility of a 2X2 magic square.

Impossibility of a 2X2 magic square

A magic square of order 2 does not exist. This can be proved by the following contradiction method. Suppose such a 2X2 square does exist, then we should have a magic square of the form.

X_1	X_2
X_3	X_4

Recall that the numbers in each box of the grid must be distinct values from 1 to 4 and that the sum of the rows and columns and diagonals must be the same.

Therefore,

$x_1+x_2 = x_3+x_4 = x_1+x_3 = x_2+x_4 = x_1+x_4 = x_2+x_3$

Then $x_1+x_2 = x_1+x_3$ which implies $x_2 = x_3$

or $x_3+x_4 = x_2+x_3$, which implies $x_2 = x_4$

Either way the numbers are not distinct. Therefore a 2X2 magic square does not exist. So magic square starts from order three only.

Position of the numbers in 3X3 magic square.

Centre square

Can 6 be placed in the centre? The answer is no. This is because that will leave no place for 9 (6 and 9 already add to 15). The numbers 7,8 and 9 can be eliminated from the centre for the same reason.

What about 4? That will not do either because there will be no place to put 1 and still come up with a sum of 15. All the numbers 1-4 can be shown to be too small for the same reason. That leaves only the number 5 in the centre square.

-	-	-
-	5	-
-	-	-

What about the rest of the numbers? They do not fall into the place easily.

9 in the corner? Why are we looking at this question? Let us see. Try 9 in the corner.

9	?	?
?	5	?
?	?	1

It can be seen that a number in the corner (in this case 9) must be combined with 6 other numbers in 3 different ways (rows, columns and diagonals) to sum to 15.

There will be only 2 blank squares that the 9 does not combine with to form a sum. We know that 9 cannot be combined with any of 6 , 7 or 8 because the sum would be greater than the magic sum of 15. Since there is no way to force all 3 of 6, 7 and 8 in the 2 blank squares, we have no choice but to conclude that the 9 cannot go to the corner.

9 on the side

4 corners are eliminated. You have no choice but to place 9 in one of the 4 side positions. It does not really matter which one you choose because rotating the square will give you the same solution. After placing 9, the number 1 also falls into place.

x	9	y
-	5	-
-	1	-

We already know that the number 6, 7 and 8 cannot be in the same row with 9. That leaves only 2,3 and 4 as the candidates for the squares. See x and y in the diagram.

If we were to place 3 in one of them, we need to place the 3 in the other to arrive at the magic sum of 15 because we can use each number only once, this eliminates 3 from being placed in the same row with 9. That leaves only 2 and 4.

4	9	2
-	5	-
-	1	-

Now there is finally enough information to begin filling in the remaining squares starting with the 2 diagonals.

One diagonal already contains 4 and 5, so the third value must be 6. The other diagonal containing 2and 5 can be completed only with 8.

4	9	2
-	5	-
8	1	6

14

So far, we have not used 3 and 7. They neatly fall into place for the final solution.

4	9	2
3	5	7
8	1	6

Magic sum or magic constant of a 3X3 magic square.

Let us take the numbers from 1 to 9. The highest number is $9=3^2$. For nXn magic square, it is n^2.

Find out the sum of the first k integers. $1+2+3+.... +k= k(k+1)/2$.

Now let $k=n^2$ which is the highest value in the sequence i.e. $n^2(n^2+1)/2$

Putting n=3, 9x10/2=45

This is the sum of all the numbers1, 2, ...9 which are used in the magic square.

The value of each row, column or diagonal is found by dividing this number by the number of rows or columns.

$(n^2(n^2+1)/2)/n = n^2(n^2+1)/2n = n(n^2+1)/2$

This is the sum of all numbers in a given row, column or diagonal in an nXn magic square. For a 3X3 magic square this works out to 15, which is called the magic sum or magic constant of the square.

The magic constant of a magic square depends on

      N- order of the square

      A- starting integer

      D- difference between successive numbers.

The magic sum $S= N (2A+D (N^2-1)) /2$

This equation can be transposed so that it is possible to calculate the starting integer A given the magic constant, difference and order.

$A= (S -DK) /N$

where $K= (N(N^2-1))/2$

15

Suppose we take a magic constant as 903. Let us work out the starting number.

Here, S = 903; N= 3; D = 1

903-(1x12) /3=297

So, the magic square is:

304	297	302
299	301	303
300	305	298

If we choose a magic constant of 1903, we will get a magic square with fractional numbers.

$637\frac{1}{3}$	$630\frac{1}{3}$	$635\frac{1}{3}$
$632\frac{1}{3}$	$634\frac{1}{3}$	$636\frac{1}{3}$
$633\frac{1}{3}$	$638\frac{1}{3}$	$631\frac{1}{3}$

Construction of a 3X3 magic square

Let us see a very simple method of constructing a magic square using numbers 1to 9. It consists of 8 steps. They are shown in the following figures.

The starting point is 1 in the centre of the top row. The next number is 1 up 1 right. When the number goes out of 3X3 square, it wraps around from top to bottom and from left to right. If the already placed number blocks the placing of a number, put it below the earlier number. All the steps are self-explanatory. Try to take another 9 numbers in arithmetic progression and construct another 3X3 magic square.

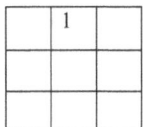 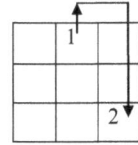 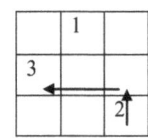 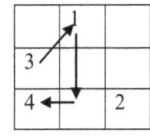

The final square is:

8	1	6
3	5	7
4	9	2

Any three sets of numbers with the following properties will form a magic square.

Each set of three numbers must be a linear sequence with the same common difference. e.g.

1, 7, 13

31, 37, 43

61, 67, 73

All of them have the common difference of 6.

The first, second and third terms of the 3 sequence must also be a linear sequence. In the above case 1, 31, 61 are with a common difference of 30. Also see, 7, 37, 67 and 13, 43, 73. The common difference is 30.

The 3X3 magic square will be

67	1	43
13	37	61
31	73	7

The magic sum =111

Let us consider 2 more examples. Note that in this case one number is negative.

27, 33, 39; -3, 3, 9; 57, 63, 69.

17

The magic square is as follows:

63	-3	39
9	33	57
27	69	3

The magic sum =99

Consider the sequence given below:

7,8,9; 13,14,15; 19,20,21

The magic square is

20	7	15
9	14	19
13	21	8

The magic sum=42.

Another easy method

To construct a 3X3 magic square there is another very simple method. Use numbers from 1to 9. Follow the steps given below.

- Write the consecutive numbers 1to 9 as shown in the figure below.
- Draw lines diagonally as shown.
- Show 3X3 grid as shown by dotted lines.
- Notice that 4 cells are empty.
- 4 numbers are lying outside the square grid.
- Use the numbers to fill the empty cell in the opposite side.

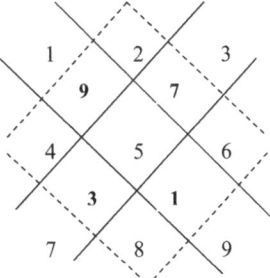

The resultant square is shown below.

4	9	2
3	5	7
8	1	6

Rotation and Reflection

There is only one 3X3 magic square. Although it is true in one sense and in another it is not. It is true because all the 3X3 magic squares are related by symmetry. Once you have one, you can get all the others by turning or flipping the one. Although we say that the 3X3 magic square has one unique solution, if we include rotation and reflection we have a total of 8 squares.

Rotation of Magic Squares

In the field of geometry and linear algebra, rotation is just a transformation that is performed by 'spinning' the object around a fixed point or in space that described the motion of a rigid body around a fixed point. This fixed point is also known as the centre of rotation. Usually rotation is done in the anticlockwise direction.

Reflection of Magic Squares

Reflection is another type of transformation. It is a transformation that is performed by 'flipping' body it is transforming. It is a mapping that transforms an object into its mirror image. e.g. the reflection of a letter p in respect to vertical line would be like q and reflection in the horizontal axis would be like b. A reflection is an involution since when applied twice in succession, every geometrical object will restore to its original state. So, a reflection of magic square is an involution. Therefore, a magic square will restore to its original form after applying reflection twice in succession. A reflection of magic square is actually isometric to a rotation of magic square.

All the 8 squares are given below.

They are all obtained by rotation and reflection of one square.

8	1	6
3	5	7
4	9	2

6	1	8
7	5	3
2	9	4

8	3	4
1	5	9
6	7	2

4	3	8
9	5	1
2	7	6

6	7	2
1	5	9
8	3	4

2	7	6
9	5	1
4	3	8

4	9	2
3	5	7
8	1	6

2	9	4
7	5	3
6	1	5

Rotation and Reflection – Sequence Designs

Although we have shown that there is only one unique solution for a 3X3 magic square, if you include rotation and reflection, we will get a total of 8 squares. All the squares exceptthe original one is called disguised squares.

In all the 8 diagrams, the sequence design has been shown below. The numbers have been joined by following manner; 1 to 2, 2to3 etc. up to 9. The number 9 is joined to number 1. This is known as the sequence design. This can be drawn for magic squares other than 3X3, like 5X5, 7X7, 9X9 etc also.

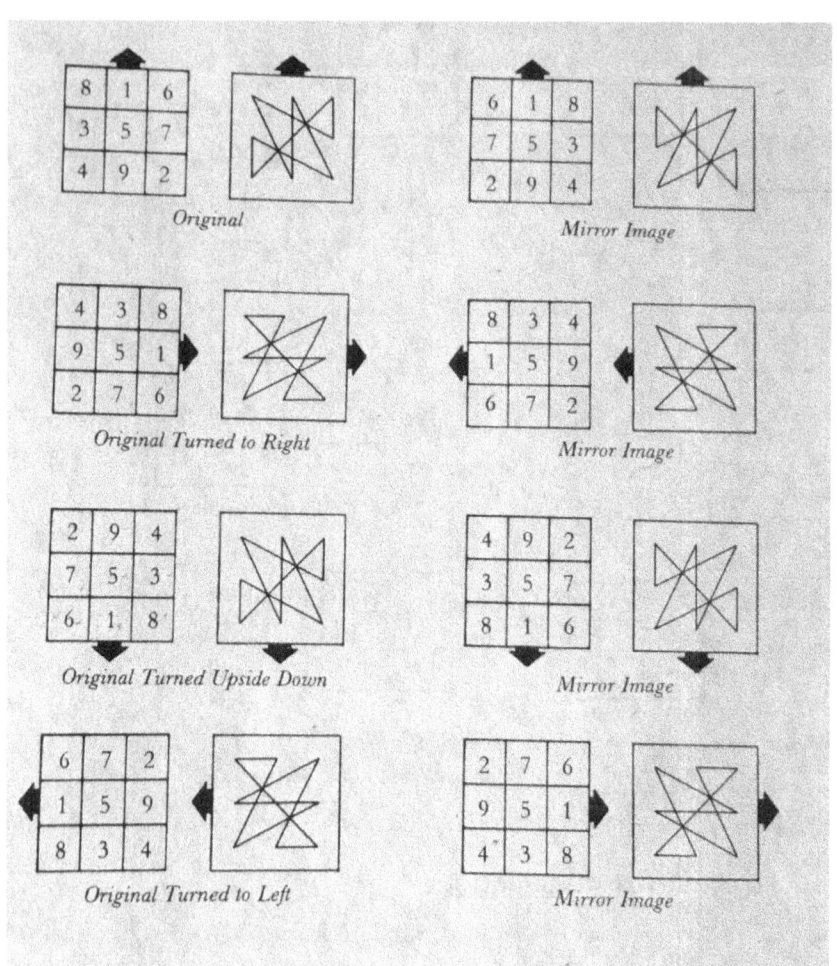

Original

Mirror Image

Original Turned to Right

Mirror Image

Original Turned Upside Down

Mirror Image

Original Turned to Left

Mirror Image

Unique Solution for 3X3 Magic Square

We have seen that excluding rotations and reflections, a 3X3 magic square has a unique solution. The rows, columns and diagonals sum to 15.

It is easy to show that no other solution is possible just by looking at all the possible ways, 3 different numbers from 1to 9 can sum to 15. There are 8 in all.

1+5+9 = 15; 1+6+8 = 15

2+4+9 = 15; 2+5+8 = 15

2+6+7 = 15; 3+4+8 = 15

3+5+7 = 15; 4+5+6 = 15

Have a close look at these 8 equations.

5 appears in 4 different expressions. 2, 4, 6 and 8each appear in 3 different expressions. 1, 3, 7 and 9 each appear in 2 different expressions. The only cell that appears in 4 sums is the central cell.

Likewise, the corner cell each appear in 3 sums and the side cells each appear in 2 sums. Hence 5 must go into the central cell, 2,4,6,8 in corner cells and 1, 3, 7, 9 inside cells.

Placing 1 in any side cell uniquely determines the placement of 9. Placing any other digit uniquely determines all the rest. A little experimentation will show that all the solutions are reflections or rotations of the one shown.

a	b	c
d	e	f
g	h	i

Let sum = j

There are 8 linear equations in 10 variables that must be satisfied by any solution to the magic square.

1. a+b+c = j
2. d+e+f = j
3. g+h+I = j
4. a+d+g = j
5. b+e+h = j
6. c+f+I = j
7. a+e+I = j
8. c+e+g = j

Number of interesting relationships can be discussed that make it easy to generate an order 3 magic square using any 3 numbers.

Let us add all the equations containing e.

$(d+e+f)+(b+e+h)+(a+e+i)+(c+e+g)=j+j+j+j$

Rearranging and combining the terms gives,

$(a+b+c)+(d+e+f)+(g+h+i)+3e = 4j$

$j+j+j+3e = 4j$

$3e = j$

The magic constant of order 3 magic square is 3 times the centre cell.

Let us take equations 6 and 7

$(a+e+i)+(c+f+i) = 2j$

$(a+c)+(e+f)+2i = 2j$

Substituting from equations 1 and 2,

$(j-b)+(j-d)+2i = 2j$

$2j-b-d+2i = 2j$

$2i = (b+d)$

$i = i/2(b+d)$

By similar reasoning, we can see that every corner cell is half the sum of the two opposite side cells. Likewise, the centre cell is half the sum of any two cells on opposite side of it. Thus,

$$e =\frac{1}{2} (a+i) =\frac{1}{2}(b+h) =\frac{1}{2}(c+g) =\frac{1}{2}(d+f)$$

Table showing all combinations with a total of 15.

	1	2	3	4	5	6	7	8	9
1+5+9=15	1				5				9
2+4+9=15		2		4					9
1+6+8=15	1					6		8	
2+5+8=15		2			5			8	
3+4+8=15			3	4				8	
2+6+7=15		2				6	7		
3+5+7=15			3		5		7		
4+5+6=15				4	5	6			
Number of combinations	2	3	2	3	4	3	2	3	2

- 5 appears in 4 combinations. Hence it will occupy the centre square because this appears on 4 lines (1 column,1 row and 2 diagonals)

- 1,3,7,9 must go to the middle of the sides because they appear only twice.
- From the above it appears that 2, 4, 6, 8 must appear in the corners. They appear 3 times (1 row, 1 column, 1 diagonal).

In each of the 4 lines that involve the centre square, the numbers are in arithmetic progression making the magic sum equal to 3 times, the centre number.

Central Value of a 3X3 Magic Square

We have seen it earlier that the central value of a 3X3 magic square with numbers from 1 to 9 is 5. Now, let us see what are the different ways in which we can find out the central value of a magic square?

- Take all the 9 numbers in the square and add them. If we use the numbers from 1to 9 the sum is 45. Divide by 9 to find out the average. It will be 5.
- Take the sum of all the four corner numbers. Find out their average.
- Take the sum of all the 4 numbers at the sides. Find out their average.
- Take the sum of any two opposite values. Take their average.
- Find out the magic sum. Take $1/3^{rd}$ of the magic sum. You will get the central value.

Look at the following square.

a	b	c
d	e	f
g	h	i

Central number 'e'

Choose 'e' to be any number between 1 to 9 inclusive of say '4' . Then the other pairs should sum to 11. Look at the following possibility.

6+5,7+(we need 4 here),8+3, 9+2, ?+1(we need 10 here)

Hence e = 4 does not work.

By letting e = 5, the other pairs should sum to 10. So we have, 6+4, 7+3, 8+2, 9+1. So e = 5 works out.

Now, the pairs of the numbers must be placed in their respective places. Look at 'a' position. There are two pairs of even numbers and two pairs of odd numbers. For each row or each column there must be either 3 odd numbers or one odd number. If an odd number is in the position 'a', there needs to be another odd number in the first row and first column. This is not possible. So, the 'a 'position has to be an even number. Thus, the pairs of even numbers must be placed on the diagonals. See the square below.

8	1	6
3	5	7
4	9	2

Now let us look at the alternative strategy. Note that there are 5odd numbers and 4 even numbers in the integer from 1to 9. Notice that sum of 15 should consist of either 1 odd or 3 odd numbers. Thus, an odd number has to be at the centre. The setup of the pattern would be as follows.

O	E	O
E	O	E
O	E	O

Let us take numbers 2,3,4,5,6,7,8,9,10.

In this case 6 has to be in the centre. The magic square will be given below.

9	2	7
4	6	8
5	10	3

The magic sum of this square is 18.

Let us take another example. Consider the number from 0to 9. In this case 4 will be in the centre. The square will be as follows.

7	0	5
2	4	6
3	8	1

The magic sum of this square is 12.

Corner Numbers in the Magic Square.

Every corner number in a magic square is half the sum of the two squares which are 'knights move' away.

Let us look at the following square.

14	42	13
22	23	24
33	4	32

In this square let us look at the 4-corner numbers.

$14 = (24+4)/2$

$13 = (22+4)/2$

$33 = 42+24/2$

$32 = (22+42)/2$

In the above magic square, the 9 numbers form 8 arithmetic progressions

2- middle row and middle column

2- main diagonals

4-broken diagonals

If all the 9 numbers form a single arithmetic progression, then the magic square can be derived from the basic magic square.

8	1	6
3	5	7
4	9	2

By a linear transformation say, Ax +B where A and B are constants and x is the value in the square. Taking A = 3 and B = 2 (i.e. 3x+2), the basic magic square transforms to

26	5	20
11	17	23
14	29	8

The numbers in this square (5,8,11,14,17,20,23,24,27) form arithmetic progressions with a common difference of 3.

Look at the following transformations.

- Take any conventional 3X3 magic square, and multiply all its entries by 2 and subtract 1 from all of them. You will obtain a magic square whose entries are the first 9 odd integers.
- Take any conventional magic square; multiply all its entries by 2. You will obtain a magic square whose entries are the first 9 even integers.

3X3 Square with Centre Number Larger than 5.

Centre number =6   There are 3 solutions.

10	1	7
3	6	9
5	11	2

9	1	8
5	6	7
4	11	3

9	2	7
4	6	8
5	10	3

Centre number=7   There are 4 solutions

12	1	8
3	7	11
7	15	2

11	2	8
4	8	10
6	12	3

10	2	9
6	7	8
5	12	4

10	3	8
5	7	9
6	11	4

Centre number =8   There are 7 solutions.

14	1	9
3	8	13
7	15	2

13	1	10
5	8	11
6	15	3

13	2	9
4	8	12
7	14	3

12	1	11
7	8	9
5	15	4

12	3	9
5	8	11
7	13	4

11	3	10
7	8	9
6	13	5

11	4	9
6	8	10
7	12	5

Centre number=9   There are 10 solutions.

16	1	10
3	9	15
8	17	2

15	1	11
5	9	13
7	17	3

15	2	10
4	9	14
8	16	3

14	2	12
7	9	11
6	17	4

14	2	11
6	9	12
7	16	4

14	3	10
5	9	13
8	15	4

13	2	12
8	9	10
6	16	5

13	4	10
6	9	12
8	14	5

12	4	11
8	9	10
7	14	8

12	5	10
7	9	11
8	13	6

Note that the number of solutions that have the centre value of 'N+1' is equal to the number of solutions that have a centre value of 'N' plus the number of new solutions that have a single digit of '1' in the middle of the top row.

See the explanation given below.

N = 5(1 solution) +2 = 3 (number '1' in the centre of the top row)
N = 6   3 solutions +1 = 4
N = 7   4 solutions +3 = 7
N = 8   8 solutions +2 =10

See the table given below:

Centre cell	Magic sum	Number of solutions
5	15	1
6	18	3
7	21	4
8	24	7
9	27	10
10	30	13
...	..	..
...	..	..

Can you fill this table up to centre cell of 20? Can you write a computer program to generate the solution?

Compliment of Magic Squares

A compliment of magic square is obtained when each number in the cell of the magic square is subtracted from the largest number of the pattern plus 1. For normal 3X3 magic square, the largest number is 9 which is 3^2. Therefore, we have to subtract every cell from 10.

Let us look at one example.

4	9	2
3	5	7
8	1	6

10-4	10-9	10-2
10-3	10-5	10-7
10-8	10-1	10-6

=

6	1	8
7	5	3
2	9	4

Compliment of a magic square

Look at the following.

8	1	6
3	5	7
4	9	2

becomes

2	9	4
7	5	3
6	1	8

8	3	4
1	5	9
6	7	2

becomes

2	7	6
9	5	1
4	3	8

Justify why the subtraction of entries from n^2+1 produces another magic square. Consider any row, column or diagonal with entries.

$x_1, x_2, x_3, \ldots x_n$

Then, the sum of the entries in the complimentary square is

$$\sum_{i=0}^{n} (n^2+1) - x_i = n\,(n^2+1) - \sum_{i=1}^{n} x_i$$

$$= n\,(n^2+1) - n\,(n^2+1)/2$$

$$= n\,(n^2+1)\,/2$$

Hence, we obtain a magic constant again and since the same argument holds for any column or main diagonal, the complimentary square is also a magic square.

Associative Magic Square

In general, for many magic square of order n, where the sum of each of its symmetric pairs is n^2+1, the square is said to be associative. 5 is the element that balances the square. Look at the following square.

4	9	2
3	5	7
8	1	6

$9+1 = 10;\ 3+7 = 10;\ 4+6 = 10;\ 2+8 = 10$

If we subtract 5 from each element, a magic square results whose magic sum is zero.

-1	4	-3
-2	0	2
3	-4	1

This is a magic square with integers from -4 to +4.

Upside down 3X3 Magic Squares

The following squares have only 3 digits 0, 1 and 2. Note that they are represented in the digital form.

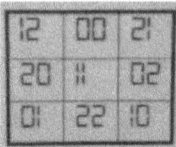

The magic constant of this square is 33.

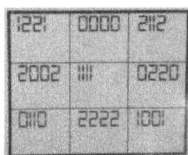

In order to have upside down we have called 110,220 as 0110, 0220 to have symmetry in the result. The magic constant of this square is 3333. Note that it is palindromic square.

Addition of 2 Magic Squares

If you add two magic squares A and B you will get C which is also a magic square. See the examples given below.

8	1	6
3	5	7
4	9	2

+

9	2	7
4	6	8
5	10	3

=

17	3	13
7	11	15
9	19	5

magic sum =15   magic sum =18   magic sum=33

You can verify the associative property of addition.

$(A+B)+C=A+(B+C)$

31

Subtraction of Two Magic Squares

Just like addition, subtraction of one magic square from another will produce another magic square. Look at the example below.

26	5	20
11	17	23
14	29	8

-

8	1	6
3	5	7
4	9	2

=

18	4	14
8	12	16
10	20	6

magic sum =51   magic sum =15   magic sum=36

Scalar Multiplication of A Magic Square

If a magic square is multiplied by a scalar quantity, the resulting square is also a magic square. See the example given below.

8	1	6
3	5	7
4	9	2

X 4   =

32	4	24
12	20	28
16	36	8

magic sum= 15   magic sum=60

Division of a Magic Square

Just as we did multiplication, we can divide all the elements of a magic square to produce another magic square. See the example given below.

8	1	6
3	5	7
4	9	2

÷ 2   =

4	1/2	3
3/2	5/2	7/2
2	9/2	1

Note that all the entries need not be the whole numbers. The magic sum also becomes half of the original sum.

An Interesting 3X3 Magic Square

The following magic square contains 9 consecutive numbers from 77 to 85.

84	77	82
79	81	83
80	85	78

It has been constructed just like any other 3X3 magic square with 9 consecutive numbers. But there is some interesting feature about the numbers in these squares.

Let us look at the centre number

$$81 = 9^2 = 3^4$$

The magic constant of this square = 3x81 = 243 = 3^5

Sum total of all numbers of the square

$$= 3x243 = 729 = 3^6$$

Add all even numbers; 80+78+82+84 = 324 = $4x3^4$

Add all odd numbers; 77+83+85+79+81 = 405 = $5x3^4$

One more interesting feature of this square is that all the numbers from 81 to729 have the same digital root.

$$81 \rightarrow 8+1 \qquad 9$$
$$243 \rightarrow 2+4+3 \qquad 9$$
$$324 \rightarrow 3+2+4 \qquad 9$$
$$405 \rightarrow 4+0+5 \qquad 9$$
$$729 \rightarrow 7+2+9 \qquad 18 \rightarrow 1+8 \qquad 9$$

Is this square not interesting?

Magic Product of 3X3 Magic Square

Let us consider one example

8	1	6
3	5	7
4	9	2

Take the products of the numbers in each row and add them up.

(8x1x6) + (3x5x7) + (4x9x2) = 48+105+72 = 225

Now do the same for the columns.

(8x3x4) + (1x5x9) + (6x7x2) = 96+45+84 = 225

Now take pair wise product in each row and add them.

(8x1) + (1x6) + (6x8) = 62

(3x5) + (5x7) + (7x3) = 71

(4x9) + (9x2) + (2x4) = 62   62+71+62 = 195

Do the same for columns

(8x3) + (3x4) + (8x4) = 68

(1x5) + (5x9) + (9x1) = 59

(6x7) + (7x2) + (2x6) = 68   68+59+68 = 195

Note that pair wise product is the same for first row and 3rdrow. Same is true for columns.

Different Sets of 9 Numbers

We find below 3 different 3X3 magic squares.

9	2	7
4	6	8
5	10	3

The magic product of the square = 465.

The pair wise product of the square = 294.

4	5	9
11	6	1
3	7	8

The magic product of the square = 414

The pair wise product of the square = 285

6	5	13
15	8	1
3	11	10

The magic product of the square = 840

The pair wise product of the square = 489

Tarry -Escott Problem and Lo Shu

Take the Lo shu square given below.

4	9	2
3	5	7
8	1	6

Read the rows as 3-digit numbers from left to right and right do left and to the following calculations.

$$492-357+816 = 951$$
$$294-753+618 = 159$$

Look at the rows and ignore the middle column to get 2-digit numbers and add them up.

$$42+37+86 = 165$$
$$68+73+24 = 165$$

Now, look at the columns and leave the middle row. Add the three 2-digit numbers.

$$84+19+62 = 165$$
$$26+91+48 = 165$$

Look at the odd digits: 1, 3, 5, 7 and 9.

Sum of the squares: $1^2+3^2+5^2+7^2+9^2=165$

There is a not too well known problem in mathematics called the Tarry-Escott problem which asks, if there are sets of integers with the same order (the same number in each set) so that the integers in each set have the same sum, the same sum of squares etc. up to and including the same sum of kth powers.

Remarkably the pattern in the Lo shu gives a solution to the Tarry-Escott problem. Start at the top left and read around the outside; the four 3-digitnumbers are

492, 276,618 and 834

Now read, them giving the other way round.

438, 816, 672 and 294.

Add up the numbers in each set,

$$492+276+618+834 = 438+816+672+294$$

$$2220 = 2220$$

Now take the squares of these numbers.

$$429^2+276^2+618^2+834^2 = 438^2+816^2+672^2+294^2$$

Now take the cubes of these numbers.

$$492^3+276^3+618^3+834^3 = 438^3+816^3+672^3+294^3$$

Another Magic and Square Property

Take the 3X3 magic square

8	1	6
3	5	7
4	9	2

Read the rows as 3-digit numbers forward and backward and square them.

$816^2+357^2+492^2 = 618^2+753^2+294^2$

What about columns?

$834^2+159^2+672^2 = 438^2+951^2+276^2$

$695556+25281+457584 = 191844+904401+76176$

$1172421 = 1172421$

Try whether it works for diagonals. Will this hold good for any 3X3 magic square?

In the 3X3 magic square example above, we have considered single digit number. If any entry contains more than 1 digit, we have to carry the extra places using techniques of linear algebra.

Let us consider the following magic square.

13	6	11
8	10	12
9	14	7

$(1300+60+11)^2+ (800+100+12)^2+ (900+140+7)^2$

$= (1100+60+13)^2+ (1200+100+8)^2+ (700+140+9)^2$

$1879641+831744+1096209 = 1375929+1710864+720801$

$3807594 = 3807594$

Now take the first 3X3 magic square.

8	1	6
3	5	7
4	9	2

Sum of the squares in the first row = sum of the square in 3rd row

$8^2+1^2+6^2 = 4^2+9^2+2^2$

$64+1+36 = 16+81+4$

$101 = 101$

We can get the same for the columns.

Sum of the squares in the first column= sum of the squares in the 3rd column

$8^2+3^2+4^2 = 6^2+7^2+2^2$

$64+9+16 = 36+49+4$

$89 = 89$

Feigenbaum Constant and Magic Constant of 3X3 Magic Square

Michell Feigenbaum noticed in 1975 that the quotient of the successive distances between bifurcation events tends to 4.66206.... This is known as Feigenbaum constant.

Now let us look at the following arithmetic progression.

1, 1.125, 1.25, 1.4375, 1.5625, 1.6875, 1.895, 2, 2.125

A3X3 magic square formed with these numbers is as follows.

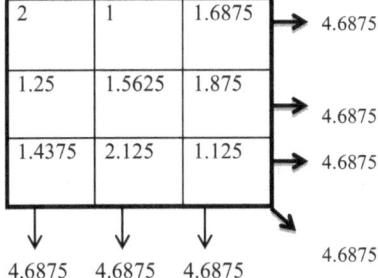

The magic constant of this square is close to the Feigenbaum constant.

Chapter 3

DUPLICATE DIGITS ALLOWED!

It is well known that in a magic square of any order, duplicate digits are not allowed. In a 3X3 square, we use a total of 9 digits and repetition of any digit is not allowed. We have proved earlier that if repetition of a number is not allowed, then there is only one unique solution for a 3X3 magic square. Of course, if you allow rotations and reflections, then we get 8 solutions.

In this chapter, let us see what happens if we allow repetitions of digits. Let us list all possible 3X3 magic squares whose elements are the single digits 1 to 9, but with duplicate digits allowed.

While the sums of three digits from 1 to 9 range between3 and 27, we find that only sums that are multiples of 3 are associated with magic squares. We will now show that there are 35 distinct 3x3 magic squares with duplicate digits being allowed.

Sum =3   1 distinct magic square

1	1	1
1	1	1
1	1	1

Sum = 6   2 distinct magic squares

1	3	2
3	2	1
2	1	3

2	2	2
2	2	2
2	2	2

Sum =9   4 distinct magic squares

1	5	3
5	3	1
3	1	5

2	3	4
5	3	1
2	3	4

2	4	3
4	3	2
3	2	4

3	3	3
3	3	3
3	3	3

Sum = 12   6 distinct magic squares

1	7	4
7	4	1
4	1	7

2	5	5
7	4	1
3	3	6

2	6	4
6	4	2
4	2	6

3	4	5
6	4	2
3	4	5

3	5	4
5	4	3
4	3	5

4	4	4
4	4	4
4	4	4

Sum = 15   9distinct magic squares

1	9	5
9	5	1
5	1	9

2	7	6
9	5	1
4	3	8

2	8	5
8	5	2
5	2	8

3	5	7
9	5	1
3	5	7

3	6	6
8	5	2
4	4	7

3	7	5
7	5	3
5	3	7

4	6	5
7	5	3
3	5	7

4	6	5
6	5	4
5	4	6

5	5	5
5	5	5
5	5	5

The second square in the first row is the classic 3X3 magic square.

Sum = 18   6 distinct magic squares

3	9	6
9	6	3
6	3	9

4	7	7
9	6	3
5	5	8

4	8	6
8	6	4
6	4	8

5	6	7
8	6	4
5	6	7

5	7	6
7	6	5
6	5	7

6	6	6
6	6	6
6	6	6

Sum = 21   4 distinct magic squares

5	9	7
9	7	5
7	5	9

6	7	8
9	7	5
6	7	8

6	8	7
8	7	6
7	6	8

7	7	7
7	7	7
7	7	7

Sum = 24   2 distinct magic squares

7	9	8
9	8	7
8	7	9

8	8	8
8	8	8
8	8	8

Sum = 27   Only 1 distinct magic square

9	9	9
9	9	9
9	9	9

In all the above cases, squares that are related by reflection about a horizontal, vertical or diagonal axis are not considered as distinct. Such problems can be solved by the method of exhaustion, using a computer.

Chapter 4

FRACTIONS AND DECIMALS

A fraction is a number that indicates a part of a unit or a part of a quantity. Fractions are written in the form of $\frac{a}{b}$ where a and b are whole numbers and the number b is not zero.

Proper fraction- numerator is less than the denominator e.g. $\frac{4}{5}$

Improper fraction- numerator is > denominator e.g. $\frac{13}{5}$

Mixed fraction- whole number and fraction e.g. $1\frac{5}{12} = 1 + \frac{5}{12}$

Equivalent fraction- different fractions which express the same amount e.g. ½ is equal to 2/4 because ½ = 2/4 = 0.5

Additive Magic Squares using Fractions:

Look at the following magic squares

$2\frac{2}{3}$	$\frac{1}{3}$	2
1	$1\frac{2}{3}$	$2\frac{1}{3}$
$1\frac{1}{3}$	3	$\frac{2}{3}$

$\frac{4}{5}$	$\frac{1}{10}$	$\frac{3}{5}$
$\frac{3}{10}$	$\frac{1}{2}$	$\frac{7}{10}$
$\frac{2}{5}$	$\frac{9}{10}$	$\frac{1}{5}$

$1\frac{3}{8}$	$\frac{1}{2}$	$1\frac{1}{8}$
$\frac{3}{4}$	1	$1\frac{1}{4}$
$\frac{7}{8}$	$1\frac{1}{2}$	$\frac{5}{8}$

$4\frac{5}{6}$	$1\frac{1}{3}$	$3\frac{5}{6}$
$2\frac{1}{3}$	$3\frac{1}{3}$	$4\frac{1}{3}$
$2\frac{5}{6}$	$5\frac{1}{3}$	$1\frac{5}{6}$

All the above 4 squares are additive magic squares using fractions.

Complete sequence of numbers and construct the magic square.

1. $2,-,-,-,-,3\frac{1}{4}, 3\frac{1}{2}, -, -$
2. $1\frac{1}{2}, -, -, -, -, 3\frac{3}{8}, 3\frac{3}{4}, -, 4\frac{1}{2}$
3. $-, -, \frac{2}{3}, 2\frac{1}{2}, -, -, -, 3\frac{1}{6}, -$
4. $-, -, -, -, -, -, 1\frac{1}{6}, 1\frac{1}{4}, -$

Arrange the following numbers into an additive magic square with a magic sum of 1.

$$\frac{1}{15}, \frac{2}{15}, \frac{4}{15}, \frac{7}{15}, \frac{8}{15}, \frac{1}{5}, \frac{2}{5}, \frac{3}{5}, \frac{1}{3}$$

In a similar manner, arrange the following numbers into a magic square with a magic sum of $\frac{1}{2}$

$$\frac{5}{36}, \frac{7}{36}, \frac{1}{18}, \frac{5}{18}, \frac{1}{12}, \frac{1}{9}, \frac{2}{9}, \frac{1}{6}, \frac{1}{4}$$

Use the following fractions to form a 3X3 magic square to yield a sum of $3\frac{3}{4}$

$$\frac{1}{4}, \frac{1}{2}, \frac{3}{4}, 1, 1\frac{1}{4}, 1\frac{1}{2}, 1\frac{3}{4}, 2, 2\frac{1}{4}$$

Put the following fractions into a 3X3 magic square such that every line adds up to $\frac{1}{2}$

$$\frac{5}{36}, \frac{7}{36}, \frac{1}{18}, \frac{5}{18}, 1\frac{1}{12}, \frac{1}{9}, \frac{2}{9}, \frac{1}{6}, \frac{1}{4}$$

Complete the following magic square puzzles

-	-	3/8
1/2	3/4	1
-	-	-

-	-	-
$1\frac{1}{16}$	-	-
$1\frac{7}{12}$	$3\frac{2}{3}$	$\frac{3}{4}$

44

<table>
<tr><td>-</td><td>-</td><td>$8\frac{3}{4}$</td></tr>
<tr><td>-</td><td>-</td><td>$9\frac{1}{2}$</td></tr>
<tr><td>$7\frac{1}{4}$</td><td>-</td><td>$5\frac{3}{4}$</td></tr>
</table>

<table>
<tr><td>$1\frac{3}{5}$</td><td>$\frac{1}{5}$</td><td>-</td></tr>
<tr><td>$\frac{3}{5}$</td><td>-</td><td>-</td></tr>
<tr><td>$\frac{4}{5}$</td><td>-</td><td>-</td></tr>
</table>

Multiplicative Magic Squares using Fractions

Look at the following 2 examples

<table>
<tr><td>$1\frac{1}{2}$</td><td>$\frac{1}{8}$</td><td>$2\frac{1}{4}$</td></tr>
<tr><td>$1\frac{1}{8}$</td><td>$\frac{3}{4}$</td><td>$\frac{1}{2}$</td></tr>
<tr><td>$\frac{1}{4}$</td><td>$4\frac{1}{2}$</td><td>$\frac{3}{8}$</td></tr>
</table>

<table>
<tr><td>5</td><td>$\frac{2}{5}$</td><td>$\frac{1}{2}$</td></tr>
<tr><td>$\frac{1}{10}$</td><td>1</td><td>10</td></tr>
<tr><td>2</td><td>$2\frac{1}{2}$</td><td>$\frac{1}{5}$</td></tr>
</table>

Complete the following 3X3 multiplicative magic squares

<table>
<tr><td>$\frac{1}{2}$</td><td>-</td><td>-</td></tr>
<tr><td>-</td><td>$2\frac{1}{2}$</td><td>$\frac{1}{4}$</td></tr>
<tr><td>-</td><td>-</td><td>$12\frac{1}{2}$</td></tr>
</table>

<table>
<tr><td>$3\frac{1}{5}$</td><td>-</td><td>-</td></tr>
<tr><td>$6\frac{2}{5}$</td><td>-</td><td>-</td></tr>
<tr><td>$\frac{1}{5}$</td><td>-</td><td>$\frac{4}{5}$</td></tr>
</table>

<table>
<tr><td>-</td><td>-</td><td>$\frac{5}{8}$</td></tr>
<tr><td>-</td><td>$1\frac{1}{4}$</td><td>$12\frac{1}{2}$</td></tr>
<tr><td>$2\frac{1}{2}$</td><td>-</td><td>-</td></tr>
</table>

<table>
<tr><td>-</td><td>-</td><td>$16\frac{2}{3}$</td></tr>
<tr><td>$8\frac{1}{3}$</td><td>$3\frac{1}{3}$</td><td>$1\frac{1}{3}$</td></tr>
<tr><td>-</td><td>-</td><td>-</td></tr>
</table>

Decimal numbers

A decimal number has 2 parts

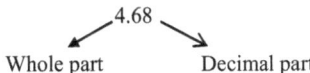

Whole part   Decimal part

You can write the decimal part of a number as a fraction.

0.23 will be 23/100

3 different types of decimals

- Exact or terminating decimal number

It is one which does not go on forever, you can write out all its digits.

e.g. 0.125

- A recurring decimal

It is a decimal number which does go on forever, but where some of the digits are repeated over and over again.

E.g. 0.12525252…….. is a recurring decimal, where '25' is repeated forever

- Irrational numbers

These are those which go on forever and do not have digits which repeat.

e.g. $\sqrt{2} = 1.4142135…..$

$\pi = 3.14159265…..$

The fact that a block of digits repeats indefinitely is denoted by a bar above the repeating digits. Look at the following examples.

$\frac{2}{3} = 0.6666……= 0.\overline{6}$

$\frac{1}{7} = 0.142857142857 = 0.\overline{142857}$

Look at the following additive magic squares

$0.8\overline{3}$	3.75	$1.\overline{6}$
$2.91\overline{6}$	$2.08\overline{3}$	1.25
2.5	$0.41\overline{6}$	$3.\overline{3}$

2.4	0.2	3.6
1.8	1.2	0.8
0.4	7.2	0.6

Solve the following additive magic squares.

2.75	1	2.25
-	2	-
-	-	1.25

9.3	3.6	-
-	6	-
-	-	2.7

The following 3 examples are multiplicative magic squares

2.4	0.2	3.6
1.8	1.2	0.8
0.4	7.2	0.6

5.4	0.3	3.6
1.2	1.8	2.7
0.9	10.8	0.6

28.8	0.6	21.6
5.4	7.2	9.6
2.4	86.4	1.8

Solve the following multiplicative magic squares

-	-	5.0
-	2.5	6.25
1.25	-	-

0.3	-	-
-	-	-
3.0	0.15	7.5

-	-	0.4
-	-	12.8
1.6	-	0.1

2.7	-	0.3
-	-	72.9
-	-	24.3

Chapter 5

CALENDAR MAGIC SQUARES

Calendars are deeply involved with mathematics. Let us see how calendars are connected to magic squares. Take the calendar of 2017. Choose any month, say, January which is given below. We have marked a 3 X 3 which is not a magic square.

Sun	Mon	Tue	Wed	Thu	Fri	Sat
1	2	3	4	5	6	7
8	9	10	11	12	13	14
15	16	17	18	19	20	21
22	23	24	25	26	27	29
29	30	31				

For a 3X3 magic square, we know that all the 3 rows, 3 columns and 2 diagonals must add to the same sum. But, in this case only the 2 diagonals and the row and column that passes through the centre sum to the same value. So, it is not a fully magic square. We also note that the magic sum will be 3 times the centre number 9x3 = 27.

$$2 \text{ diagonals} : 1 + 9 + 17 = 27$$

$$3 + 9 + 15 = 27$$

$$2^{nd} \text{ row} \quad : 8 + 9 + 10 = 27$$

$$2^{nd} \text{column} \quad : 2 + 9 + 16 = 27$$

If we can make changes in this square so that all the rows, columns and diagonals sum to 27, then we have a fully 3X3 magic square. Let us add all the 9 numbers in this square.

$$1+2+3+8+9+10+15+16+17 = 81$$

This equals 9 times the number in the centre. 9x9 = 81. Because 9 is the centre number, the magic sum of this square must be 9 x 3 = 27.

Note that the numbers marked in the calendar form 3 arithmetic progressions.

1,2,3; 8,9,10; 15,16,17.

The common difference in this case is 1. If you look at the columns (1,8,15; 2,9,16; 3,10,17) the common difference in the arithmetic progressions is 7.

The mean value of the nine numbers is 9. Therefore this will occupy the centre of the 3X3 magic square. The other 8 numbers can be positioned from knowledge of the construction of 3X3 magic squares. It is given below:

16	1	10	→ 27
3	9	15	→ 27
8	17	2	→ 27

27   27   27   27   27

Let us take a look at the following pattern. Change the following square in to a magic square.

1	2	3
4	5	6
7	8	9

- Take the diagonals 1,5,9 and 3,5,7 to the middle column and middle row cells
- Take the middle column and middle row cells to diagonal cells in the reverse order

	1	
3	5	7
	9	

8	1	6
3	5	7
4	9	2

Can you change a 3rd order incomplete magic square from a calendar into a complete magic square by this method? Let us try.

1	2	3
8	9	10
15	16	17

- Take 1,9,17 and 3,9,15 and put it in middle column and middle row.
- Take 2,9,16 and 8,9,10 and shift into diagonals in the reverse order.

16	1	10
3	9	15
8	17	2

The magic sum of this square is 3x9 = 27.

Is there any more magical property in this square? Square the numbers in each cell. We will get the following square.

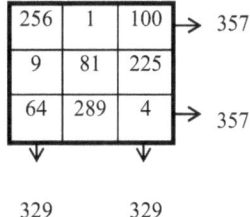

End row sums and end row columns are equal. Middle row, middle column and 2 diagonal sums are not equal.

You may be wondering what is the magical property we have discovered. This solves a puzzle of following numbers that can be expressed as a sum of 3 squares in 2 different ways!

$$8^2 + 17^2 + 2^2 = 357 \qquad 16^2 + 3^2 + 8^2 = 329$$

$$16^2 + 1^2 + 10^2 = 357 \qquad 10^2 + 15^2 + 2^2 = 329$$

Some Tricks with Calendar Squares

Take any 3x3 grid from a calendar. You can tell the sum of all the 9 numbers if you know the central number. Multiply the central number by 9 to get the answer. If you are given the lowest number in the grid, add 8 to that number and then multiply by 9 to get the sum of all 9 numbers.

If you are given a 5x4 grid and asked to give the sum of all 20 numbers in the grid, take the smallest and largest number in the grid. Add them and multiply by 10 to get the sum of 20 numbers in the grid.

1. Can you explain why the 5x4 grids work?
2. Instead of having the natural numbers from 1 to 31 as depicted in the calendar, we would have integers in arithmetic progression (the first term is 'a' and the common difference is $d \neq 1$), would the 2 tricks for 3X3 and 5X4 grids still work? Will it work with a geometric progression or some other sequence, say, the square numbers or the Fibonacci sequence?

Chapter 6

ALPHAMAGIC SQUARES

It is a magic square in which the number of letters in the name of each number in the square generates another magic square. The alphamagic squares are language dependent. These were invented by Lee Sallows in 1986.

Let us consider one example in English language.

5	22	18
28	15	2
12	8	25

The magic sum of this 3X3 magic square is 45. Now let us write these numbers in words.

FIVE	TWENTY TWO	EIGHTEEN
TWENTY EIGHT	FIFTEEN	TWO
TWELVE	EIGHT	TWENTY FIVE

Now, counting the number of letters in each number would generate the following square which also turns out to be magic.

4	9	8
11	7	3
6	5	10

So, the original square is called alphamagic. Did you notice any other special property of the 9 numbers in the above square? They are all consecutive numbers from 3 to 11. The magic sum of this magic square is 21.

Russian Alphamagic Square

74	50	92
90	72	54
52	94	70

- Each row, column and diagonal adds to 216.
- It is called as Russian alphamagic square because the number of letters in each Russian number creates another magic square in which each row, column and diagonal adds to 36.
- This is the first Cyrillic (Russian) alphamagic square to be documented.

семьдесят четыре (15)	пятьдесят (9)	девяносто два (12)	= 36
девяносто (9)	семьдесят два (12)	пятьдесят четыре (15)	= 36
пятьдесят два (12)	девяносто четыре (15)	семьдесят (9)	= 36
= 36	= 36	= 36	= 36

15	9	12
9	12	15
12	15	9

The magic sum of this square is 36. Such squares are available in French, German and Latin also.

Few English alphamagic squares of order 3 are given below.

1.

8	12	18
25	15	5
12	11	22

EIGHT	TWELVE	EIGHTEEN
TWENTY FIVE	FIFTEEN	FIVE
TWELVE	ELEVEN	TWENTY TWO

5	8	8
10	7	4
6	6	9

2.

115	72	48
78	45	12
42	18	75

ONE HUNDRED FIFTEEN	SEVENTY TWO	FORTY EIGHT
SEVENTY EIGHT	FORTY FIVE	TWELVE
FORTY TWO	EIGHTEEN	SEVENTY FIVE

17	10	10
12	9	6
8	8	11

3.

18	69	48
75	45	15
42	31	72

EIGHTEEN	SIXTY NINE	FORTY EIGHT
SEVENTY FIVE	FORTY FIVE	FIFTEEN
FORTY TWO	THIRTY ONE	SEVENTY TWO

8	9	10
11	9	7
8	9	10

4.

4	101	57
107	54	1
51	7	104

FOUR	ONE HUNDRE D ONE	FIFTY SEVEN
ONE HUNDRE D SEVEN	FIFTY FOUR	ONE
FIFTY ONE	SEVEN	ONE HUNDRE D FOUR

4	3	10
15	9	3
8	5	14

5.

5	102	58
106	55	2
52	8	105

FIVE	ONE HUNDRED TWO	FIFTY EIGHT
ONE HUNDRED SIX	FIFTY FIVE	TWO
FIFTY TWO	EIGHT	ONE HUNDRED FIVE

4	13	10
15	9	3
8	5	14

Some Mathematical Operations with Alphamagic Squares

Consider the following squares.

5	22	18
28	15	2
12	8	25

4	9	8
11	7	3
6	5	10

Add 100 to each cell of the above 2 magic squares. Then add the corresponding cells together to make a new magic square.

105	122	118
128	115	102
112	108	125

\+

104	109	108
111	107	103
106	105	110

=

209	231	226
239	222	205
218	213	235

The magic constant of the final square is 222 x 3 = 666. This is known as the Beast number. It may be noted that '666' shows up twice in the Bible. The phraseology is "six hundred three score and six".

Let us consider the alphamagic square, we started with in the beginning of this chapter.

5	22	18
28	15	2
12	8	25

Let us add 100 to each cell in the above square

105	122	118
128	115	102
112	108	125

Expressed in words and counting the letters this becomes

14	19	18
21	17	13
16	15	20

The magic sum of this square is 51

Instead of adding 100 to each cell, what happens if we add 200? See the result below.

205	222	218
228	215	202
212	208	225

Expressed in words and counting the letters, this becomes

14	19	18
21	17	13
16	15	20

Before ending this chapter, let us see 2 more examples

249	320	328
378	299	220
270	278	349

The magic sum of this square is 897. There are 3 arithmetic sequences. 220, 249, 278; 270, 299, 328; 320, 349, 378. Note that the common difference is 29. From the above square we can generate the following square.

19	18	23
24	20	16
17	22	21

The magic sum is 60. Look at the 3 arithmetic progressions. The common difference is 3.

Look at the last example.

215	372	298
378	295	212
292	218	375

From this square, we can generate the following square

17	22	21
24	20	16
19	18	23

Note that like the previous example, here also the common difference is 3. It may also be noted that both the squares contain consecutive numbers from 16 to 24.

Chapter 7

RAMANUJAN AND MAGIC SQUARES

Srinivasa Ramanujan, an Indian mathematician who was labeled as 'the man who knew infinity' was a mathematical genius in the 20th century. He was born on December 22, 1887 in Erode, a town in South India.

Even as a school boy he has constructed magic squares of different dimensions up to 8X8. He designed the following formula for constructing a 3X3 magic square.

C+Q	A+P	B+R
A+R	B+Q	C+P
B+P	C+R	A+Q

A, B and C are integers in arithmetic progression and so are P, Q and R.

Let us try using the values of 2,4,6 for A,B,C and 7, 10, 13 for P, Q, R. We get the following 3X3 magic square.

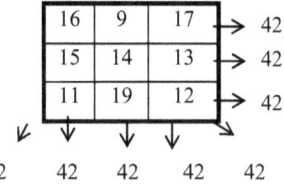

We find that the magic sum of the square is 42. All the rows, columns and diagonals sum to the same value.

If A,B ,C and P,Q,R are not in arithmetic progression, the rows and columns may add to the same sum, but not the diagonals. Choose the values 1,5,8 for A,B,C and 2,4,7 for P,Q,R. You end up with the following square.

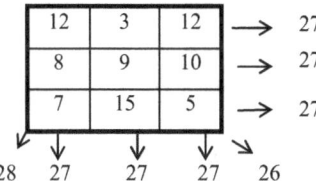

The rows and columns add up to 27 but the diagonals add up to 26 and 28. Also note that the number 12 is repeated.

Chapter 8

ROBERTSON SQUARES

Let us start with the famous 3X3 magic square.

2	7	6
9	5	1
4	3	8

We have already seen that the sum of the 3 rows, 3 columns and 2 diagonals gives us the same sum i.e. 15. Now let us see what will happen if instead of adding the digits we multiply them.

1st row product = 2 x 7 x 6 = 84

2nd row product = 9 x 5 x 1 = 45

3rd row product = 4 x 3 x 8 = 96

225

If we add all the 3 products, we get 225 which is the square of the magic constant (15) of the 3X3 square.
Now, let us look at the sum of column products.

1st column product = 2 x 9 x 4 = 72

2nd column product = 7 x 5 x 3 = 105

3rd column product = 6 x 1 x 8 = 48

225

That sum is also 225. It should be noted here that after rotation and reflection also, this equality will hold.

Let us see what will happen if we take the sum of the diagonal products.

2 x 5 x 8 = 80

6 x 5 x 4 = 120

200

This total is 200.

Martin Gardner had a doubt in his mind. He wanted to know whether there are any order 3 magic squares for which the sum of the products of 2 diagonals is also the same as the other 2 sums. However, he was unable to construct such a square. He corresponded with Robertson, who sent a proof that there is an infinity of such squares.

Let us see how to construct such squares. Look at the algebraic structure of order - 3 magic square.

a+b	a-b-c	a+c
a-b+c	a	a+b-c
a-c	a+b+c	a-b

Start with any sequence of square numbers in arithmetic progression x^2, y^2 and z^2. For 'a' in the algebraic matrix, substitute the value of '2y'. For 'b' substitute the value of 'x' and for 'c' substitute the value of 'z'.

Let us take the simplest example. 1^2, 5^2, 7^2. The 3 squares are 1, 25 and 49. They are in arithmetic progression with a difference of 24. Within the matrix, then 'a' has a value of 2 x 5 = 10, 'b' has a value of 1 and 'c' has a value of 7. The result is a magic square shown below.

11	2	17
16	10	4
3	18	9

Sum of row products $= (11 \times 2 \times 17) + (16 \times 10 \times 4) + (3 \times 18 \times 9)$

$$= 374 + 640 + 486 = 1500$$

Sum of column products $= (11 \times 6 \times 3) + (2 \times 10 \times 18) + (17 \times 4 \times 9)$

$$= 528 + 360 + 612 = 1500$$

Sum of diagonal products $= (11 \times 10 \times 9) + (17 \times 10 \times 3)$

$$= \ 990 + 510 \ = \ 1500$$

Thus, we find that the sum of row products, column products and diagonal products are the same.

These squares produced by Robertson are shown below. Above each square is shown the arithmetic progression of the square numbers which generate the square and the values taken by a, b and c in the matrix. Below each square is its magic product- addition sum.

1) 7^2 , 13^2, 17^2

a = 26, b = 7, c =17

33	2	43
36	26	16
9	50	19

Addition of magic products = 26, 364

2) 7^2 , 17^2, 23^2

a = 34, b = 7, c = 23

41	4	57
50	34	18
11	64	27

Addition of magic products = 58, 956

3) 17^2 , 25^2, 31^2

                    a $=$ 50, b $=$ 17, c $=$ 31

67	2	81
64	50	36
19	98	33

Addition of magic products = 187,500

Let us call these squares Robertson squares. They are not only magic in the traditional sense, but are also magic in an entirely different sense. We know that the magic constant of any 3X3 magic square is 3a, which is 3 times the central number. Robertson squares have in addition what we call the multiplicative, additive constant. It is a($2a^2$- b^2 - c^2) or more simply $3a^2/2$.

A simplified account of how Robertson discovered the procedure for constructing such squares is given below.

From the algebraic matrix, it is easy to determine that the sum of row (or column) products is a ($3a^2$ - $3b^2$ - $3c^2$). The sum of the diagonal products is a ($2a^2$ - b^2 - c^2).

Setting these equal and simplifying yields $a^2 = 2(b^2 + c^2)$.

Because 'a' must be even, we can rewrite the equation as 2 $(a/2)^2 = b^2 + c^2$. This expression tells us that b^2, $(a/c)^2$ and c^2 are squares in arithmetical progression. There is an infinite number of such triples and well known formulas for producing them. This makes it easy to construct an infinite number of Robertson squares.

It is to be noted that at the most 3 squares can be in arithmetic progression. Such triples are closely related to Pythagorean triangles – right triangles with integral sides. The smallest square root of a number in the progression is the difference between triangle's legs, the largest square root is the sum of the legs and the middle square root is the Pythagorean triangle's hypotenuse.

Consider e.g. the familiar 3,4,5 Pythagorean triangle. The square of these numbers are 1, 25 and 49. They form the Pythagorean triangle that generates the smallest Robertson square. The difference between its legs is 7

The 3 squares shown correspond to Pythagorean triangles with sides 5, 12, 13; 8, 15, 17; and 7, 24, 25. Thus from any Pythagorean triangle you can construct a Robertson magic square.

A question remains. Are there order – 3 squares other than the lo shu such that the sum of the row products equals the sum of column products and such that this sum also equals the square of the magic constant (diagonals not considered) ?

Robertson has shown that the lo shu is the only 3X3 square in distinct positive integers that has this property. If zero is allowed, the square given below can meet this proviso.

14	0	10
4	8	12
6	16	2

Sum of row products = 576   Sum of column products = 576

This sum is also the square of the magic constant of this square. $24^2 = 576$.

Similar squares exist if negative numbers are allowed. Lee Sallows has found the following square.

1	-4	3
2	0	-2
-3	4	-1

The sum of row products, sum of column products and sum of diagonal products in each case is zero. It is to be noted that zero is also the square of the magic constant. Note that adding 5 to each number produces lo shu.

An interesting property of the order – 3 magic squares long known, is that the sum of the squares of the first row equals the sum of the squares of the third row. This is also true of course of the sums of the squares in the first and third columns, though the two sums are never the same. Roberson has found a simple way to construct order – 3 magic squares with distinct positive integers such that the sum of the row (or column) products is an integral multiple of the square of the magic constant.

Chapter 9

MAGIC SQUARE OF SQUARES

We cannot construct any 3X3 magic square composed of 9 squared numbers. But, we can construct 3X3 magic squares with less than 9 squared numbers. If we do not impose the diagonal restrictions, then we can construct 3X3 magic squares with 9 squared numbers.

One diagonal fails

113^2	2^2	94^2
82^2	74^2	97^2
46^2	127^2	58^2

Evaluating the squares we get:

12,769	4	**8,836**
6,724	**5,476**	9,409
2,116	16,129	3,364

The magic sum of this square is $21,609=147^2$. The diagonal marked fails.

35^2	3495^2	2958^2
364^2	2125^2	1785^2
2775^2	2058^2	3005^2

Evaluating the squares, we get:

1225	12,215,025	8,749,764
13,264,164	**4,515,625**	3,186,225
7,700,625	4,235,364	**9,030,025**

The magic sum of this square is 20,966,014. It is not a perfect square since the diagonal marked fails.

Two Diagonals Fail

4^2	23^2	52^2
32^2	44^2	17^2
47^2	28^2	16^2

Evaluating the squares we get:

16	529	**2704**
1024	**1936**	289
2209	784	**256**

The sum of rows and columns is $3249 = 57^2$. In this case both the diagonals fail.

3X3 Magic Square using 7 Distinct Squared Integers

373^2	289^2	565^2
360,721	425^2	23^2
205^2	527^2	222,121

Evaluating the squares we get:

139,129	83,521	319,225
360,721	180,625	529
42,025	277,729	222,121

The magic sum of this square is 541,875.

This is the only known example of a 3X3 magic square with 7 distinct squared integers.

3X3 Square with 5 Perfect Squares

2521	49	1465
289	1345	2401
1225	2641	169

The perfect squares are marked. The magic sum of this square is 4035. Three rows, three columns and two diagonals give the same sum. Earlier we have considered the following square with 7 squared numbers.

373^2	289^2	565^2
360,721	425^2	23^2
205^2	527^2	222121

We can easily transform it, each cell x becoming (x-1)/8.

The triangular number $T_n = n(n+1)/2$

For example: $23^2 = x$

$$(x-1)/8 = (529-1)/8 = 66 = T_{11}$$

This is the 11th number in triangular number series.

T_{186}	T_{144}	T_{282}
45090	T_{122}	T_{11}
T_{102}	T_{263}	27,765

Look at the following 2 magic squares with the same magic sum and 7 numbers in common.

3^2	447	219
T_{29}	15^2	T_5
T_{21}	T_2	21^2

T_5	21^2	219
429	15^2	T_6
T_{21}	3^2	T_{29}

3X3 Magic Squares consisting of Primes

It is impossible to construct a 3X3 magic square consisting of 9 squared primes. We can have such squares with 5 and 6 squared primes.

65,737	31^2	223^2
151^2	197^2	54817
167^2	76,657	109^2

19^2	7^2	313
193	241	17^2
13^2	433	11^2

In the above square 313,193, 241 and 433 are primes. Of these 241 is a twice prime. It is also a lucky number. So, it is a lucky prime. It is also a Proth prime.

Look at the following square constructed by Lee C.F Sallows

127^2	46^2	58^2
2^2	113^2	94^2
74^2	82^2	97^2

The resultant values are given below.

16129	2116	3364
4	12,769	8836
5476	6724	9409

The rows, columns and one diagonal (top right to bottom left) give the same sum of 21,609. The other diagonal (top left to bottom right) does not give the same sum.

Kevin Brown has constructed the following square.

4^2	23^2	52^2
32^2	44^2	17^2
47^2	28^2	16^2

The resultant values are given below:

16	529	2704
1024	1936	289
2209	784	256

Three rows and three columns give the same sum i.e. $3249 = 57^2$. Both the diagonals do not give the same sum.

A square of this kind constructed using prime number squares is given below.

11^2	23^2	71^2
61^2	41^2	17^2
43^2	59^2	19^2

The resultant values are given below:

121	529	5041
3721	1681	289
1849	3481	361

Three rows and three columns give the same sum i.e. 5691. Both the diagonals do not give the same sum.

Chapter 10

MULTIPLICATIVE MAGIC SQUARES

Just like addition magic squares, we can construct multiplication magic square also. In this case, we multiply the numbers in each row, each column and each diagonal and the product remains the same.

Look at the following square.

256	2	64
8	32	128
16	512	4

The magic product of this square is $32768 = 32^3$. It may be interesting to see how we arrived at this square.

Take the classical 3X3 addition magic square.

8	1	6
3	5	7
4	9	2

Now, raise each number to the power of 2.

2^8	2^1	2^6
2^3	2^5	2^7
2^4	2^9	2^2

Instead of raising the numbers to the power of 2, can we raise it to the power of 3 or for that matter to any power n? The answer is 'yes'. Look at the example given below.

3^8	3^1	3^6
3^3	3^5	3^7
3^4	3^9	3^2

6561	3	729
27	243	2187
81	19,683	9

The magic product of this square is $14,348,907=3^{15}$

Look at the following 9 numbers

1, 2, 3, 4, 6, 9, 12, 18 and 36

Can you form a 3X3 multiplicative magic square?

Let us rearrange the numbers as follows:

1 ,2, 4 ; 3, 6, 12 ; 9, 18, 36

We have 3 geometric progressions with a common factor of 2. Hence, we can arrange them in the form of a multiplicative magic square.

18	1	12
4	6	9
3	36	2

The magic product of this square is $216 = 6^3$

Earlier we have used the classical 3X3 addition magic square and raised the numbers to the power of 2 and 3 to produce multiplication magic squares.

Now we will try the same technique but we will use the integers 0 to 8 to form the addition magic square.

7	0	5
2	4	6
3	8	1

2^7	2^0	2^5
2^2	2^4	2^6
2^3	2^8	2^1

128	1	32
4	16	64
8	256	2

The magic product of this square is $4096 = 16^3$

We can use the same process but involving negative exponents.

4	-3	2
-1	1	3
0	5	-2

3^4	3^{-3}	3^2
3^{-1}	3^1	3^3
3^0	3^5	3^{-2}

81	$\dfrac{1}{27}$	9
$\dfrac{1}{3}$	3	27
1	243	$\dfrac{1}{9}$

The magic product of this square is $27=3^3$.

The magic squares discussed above are called multiplication magic squares. They can also be termed as geometric magic squares. They are analogous in all respects to arithmetical magic squares.

Any feature produced in an arithmetical square can likewise be produced in a geometric square, the only difference being that the features of the former are shown by summations while those of the latter are shown by products. Where we use an arithmetical series for one, we use a geometric series for the other and where one is constructed by a method of differences the other is constructed by ratios.

Because of the large numbers involved, the geometric squares may be considered unattractive. But, they are interesting to study. The actual squares may not be constructed. The absurdity of constructing large squares can be easily shown. e.g. suppose we were to construct a 8th order square using the series $2^0, 2^1, 2^2, 2^3,\ldots\ldots\ldots,2^{63}$, the lowest number would be 1 and the highest number would be 9, 223, 372, 036, 854, 775, 808. Imagine multiplying together the numbers in any of its rows or columns.

Analogous to the arithmetical squares the geometric squares maybe constructed with a straight geometric series, a broken geometric series or a series which has no regular progression.

As we have seen earlier any base can be used for constructing a geometric magic square. Using the powers of 10, we can have the following square.

10^8	10^1	10^6
10^3	10^5	10^7
10^4	10^9	10^2

Note that multiplying the numbers with the same base amounts to adding exponents.

More general version with a's and b's with a and b relatively prime, the magic product is a^3b^3

Put a = 2 and b = 3, the magic product is 216.

ab^2	1	a^{2}b
a^2	ab	b^2
b	a^2 b^2	a

18	1	12
4	6	9
3	36	2

The minimum magic product of a multiplicative magic square for order 3 is 6^3 i.e. 216.

We have earlier considered the following numbers

1, 2, 4, 8, 16, 32, 64, 128 and 256 and placed them in a 3X3 grid so that the product of each row, column and diagonal gives the same value.

Consider the above numbers as powers of 2.

$2^0 = 1, 2^1 = 2, 2^2 = 4$.........$2^8 = 256$

Let the magic product of each row, column and diagonal be P. The product of each row is equal to P^3, but it will also be equivalent to multiplying every value in the grid.

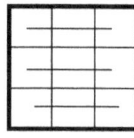

Therefore $2^0 x 2^1 x 2^2 x$.........$x 2^8 \Rightarrow 2^{36} = P^3$

$P = 2^{12} = 4096$

If we now consider the 3 values that will be found along the same row, column or diagonal as the largest value.

$P = 2^8 \times 2^a \times 2^b$ and $\quad a+b = 4$

Hence $a+b$ can be $0+4$ or $1+3$ but it cannot be $2+2$ as a and b must be different values. Because 2^8 belongs to a triplet which has 2 solutions, it must be placed in one of the edge squares (a corner would have 3 solutions and centre square would have 4 solutions).

Now we consider the smallest value in the same way.

$P = 2^0 \times 2^c \times 2^d$ and $\quad c+d = 12$

Therefore $c+d$ can be $8+4$ or $7+5$ which means that it must be placed on an edge square and it will be found opposite to 2^8. So, without loss of generality, we can fill in the values for 2^8, 2^1, 2^3, 2^4 and 2^0 (values in blue in the diagram below).

2^1	2^8	2^3
2^6	2^4	2^2
2^5	2^0	2^7

The red values can be completed by first ensuring that the sum of the indices in the 2 diagonals adds to 12, then the first and last columns can be completed. Hence, we obtain the full unique solution (considering rotations and reflections to be equivalent).

2	256	8
64	16	4
32	1	128

Some properties of multiplicative magic squares

- It is impossible to construct a multiplicative magic square using consecutive integers.
- A multiplicative magic square remains magic when all the numbers are multiplied by the same constant.
- A multiplicative magic square remains magic when all the numbers are squared, cubed or raised to the fourth power and so on.

Chapter 11

NAVAGRAHAS AND MAGIC SQUARES

Navagrahas refer to the 9 planets. This includes Mars, Mercury, Jupiter, Saturn, Venus, the Sun, the Moon as well as positions in the sky Rahu (north or ascending lunar node) and Ketu (south or descending lunar node). Of these Rahu and Ketu are actually classified as chayagrahas (meaning shadow grahas and not real ones). See the list below:

1. Surya – Sun
2. Chandra – Moon
3. Mangal – Mars
4. Budha – Mercury
5. Guru – Jupiter
6. Shukra – Venus
7. Shani – Saturn
8. Rahu ⎤
9. Ketu ⎦ Chayagrahas

As per the Hindu custom, the navagrahas are typically placed in a single square with the Sun(Surya) in the centre and other deities surrounding Surya; no two of them are positioned to face each other.

You might have seen the navagrahas laid out in consecration पतिष्ठा in the temples in India. The Sun will be in the centre.

Ketu	Brahaspati	Budha
Shani	Surya	Shukra
Rahu	Mangal	Chandra

Now substitute the numerical values given to each in table given earlier. You will get the following square.

9	5	4
7	1	6
8	3	2

There is nothing special about this square.

Instead you put Brahaspati in the centre and the other grahas as shown below.

Shukra	Surya	Rahu
Shani	Brahaspati	Mangal
Chandra	Ketu	Budha

Plug in the numbers as before and you will get the following square

6	1	8
7	5	3
2	9	4

This is the regular 3X3 magic square.

Now let us look at the 3X3 magic square for all the 9 planets starting with Sun(Surya) and ending with Ketu(chayagraha). The centre numbers of these squares range from 5 to 13. See the 9 individual squares given below.

Surya(Sun)   Numbers 1 to 9

6	1	8
7	5	3
2	9	4

Magic Sum=15

Chandra(Moon) Numbers 2 to 10

7	2	9
8	6	4
3	10	5

Magic Sum=18

Mangal(Mars) Numbers 3 to 11

8	3	10
9	7	5
4	11	6

Magic Sum=21

Budha(Mercury)   Numbers 4 to 12

9	4	11
10	8	6
5	12	7

Magic Sum=24

Brihaspati(Jupiter) Numbers 5 to 13

10	5	12
11	9	7
6	13	8

Magic Sum=27

Shukra(Venus) Numbers 6 to 14

11	6	13
12	10	8
7	14	9

Magic Sum=30

Shani(Saturn) Numbers 7 to 15

12	7	14
13	11	9
8	15	10

Magic Sum=33

Rahu(chayagraha) Numbers 8 to 16

13	8	15
14	12	10
9	16	11

Magic Sum=36

Ketu(chayagraha) Numbers 9 to 17

14	9	16
15	13	11
10	17	12

Magic Sum=39

Now let us add the magic sums of all the 9 magic squares.

15+18+21+24+27+30+33+36+39 = 243

Dividing the total by 3, we get 243/3 = 81

81 will be the magic sum, if we construct a magic square with the 9 magic sums of the 9 squares. See the figure below.

30	15	36
33	27	21
18	39	24

If we divide the individual elements of this square by 3, we get a magic square with central element as 9 and magic sum as 27.

10	5	12
11	9	7
6	13	8

We have already seen that this is the magic square for Brihaspati.

Constructing A 9X9 Magic Square of 9 Grahas

Is it possible to create such a square? Yes. We give below this square.

11	6	13	6	1	8	13	8	15
12	10	8	7	5	3	14	12	10
7	14	9	2	9	4	9	16	11
12	7	14	10	5	12	8	3	10
13	11	9	11	9	7	9	7	5
8	15	10	6	13	8	4	11	6
7	2	9	14	9	16	9	4	11
8	6	4	15	13	11	10	8	6
3	10	5	10	17	12	5	12	7

The magic constant of this square is 81. This is equivalent to the sum of the middle digits of all 3X3 squares.

5+6+7+8+9+10+11+12+13 = 81

In addition to the magic squares for like navagrahas, there is also a magic square for all the Lokpals. They are the protectors or the guardians of the world. The names of the various Gods are mentioned in the following verse.

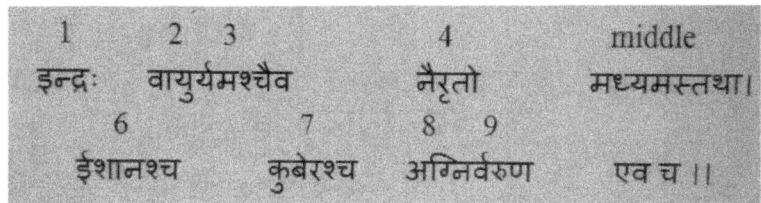

This was composed by Leelavati who was the daughter of Bhaskaracharya (Bhaskara II), the famous Indian mathematician.

Indra – 1

Vayu – 2

Yama – 3

Nairuta – 4

Ishana – 6

Kubera – 7

Agni – 8

Varuna – 9

Note that only number 5 has been left out. That belongs to the central square which is called *madyama*.

There are a total of 10 directions. If we leave out the 'above' and 'below' we have 8 directions.

North, South, East, West, Northeast, Southeast, Northwest and Southwest.

Vayu – guardian of NW

Kubera - guardian of N

Ishana - guardian of NE

Indra - guardian of E

Agni - guardian of SE

Yama - guardian of S

Nairita - guardian of SW

Varuna-guardian of W

We can form a magic square using the numbers mentioned in the verse given earlier.

Vayu	Kubera	Ishana
Varuna		Indra
Nairita	Yama	Agni

2	7	6
9	5	1
4	3	8

The magic sum is 15 and the central square is occupied by 5. The middle square has no guardian. As per the verse the number 5 is assigned to the middle square. The other 8 numbers refer to the 8 directions.

Kubera Kolam

It is a floor painting or rangoli done in South India. It is usually done with rice flour.

Kubera is regarded as the God of wealth in Hindu mythology. He is the son of Lord Brahma. He is also known as Dhanapati, the lord of wealth. Kubera is always remembered with the Goddess of fortune Lakshmi. On Diwali day, to attain prosperity, people do Lakshmi Kubera puja.

Kubera yantra or Kubera kolam is a 3X3 magic square.

27	20	25
22	24	26
23	28	21

The magic sum is 72 which boils down to 9(7+2) which is a divine number. Did you notice that this is nothing but the basic 3X3 square given below to which the number 19 has been added to each element?

8	1	6
3	5	7
4	9	2

Chapter 12

PALINDROMES

A word, verse or passage that reads the same forward or backward is palindrome. Its origin is from Greek palindromes for "running back again". Look at the following examples. Mom, eve, level, Malayalam, Madam I'm Adam, Able was I ere I saw Elba

A slightly larger one, devised by Peter Hilton, a code breaker on the British team that cracked the German enigma is

"Doc note. I dissent. A fast never prevents a fatness. I diet on cod."

A palindrome number is number that reads the same forward and backward.

161,757,12121,3535353 etc

More generally, it is symmetrical number written in base as $a_1, a_2, a_3 \ldots \ldots \ldots a_3, a_2, a_1$. In the familiar base 10 system, there are nine 2- digit palindromic numbers.

11, 22, 33, 44, 55, 66, 77, 88 and 99.

There are 90 palindromes with 3-digits.

101,111,121,131,........,959,969,979,989,999.

There are 90 palindromes with 4-digits

1001,1111,1221,......,9559,9669,9779,9889,9999 giving a total of 189 palindrome numbers below 10^4. Below 10^5 there are 1099 palindromes and for other exponents of 10^n there are

1999, 10999, 19999, 109999, 199999, 1099999....number of palindromes.

It is conjectured but has not been proven, that there are an infinite number of palindrome prime numbers. With the exception of 11, palindrome primes must have an odd number of digits.

Some examples of palindromic primes are given below.

101, 131, 151, 181, 757, 919, 91019 etc.

Many palindrome primes are also in arithmetic progression.

10301,13331,16361,19391 difference 3030

13931, 14741, 15551 difference 810

70607, 73637, 76,667, 79697 difference 3030

94349, 94649, 94949 difference 300

If you can find 9 palindromic primes in arithmetic progression, you can create an unusual magic square.

Let us take the famous 3X3 magic square.

8	1	6
3	5	7
4	9	2

How will you convert this magic square into a palindrome 3X3 magic square?

You try to think and find out the solution yourself. If you are unable to find out, then read further to get the solution.

Actually, it is very simple. In the magic square we have considered above all the 9 numbers are single digit numbers. Let us convert each into 3-digit number by placing 1 on each side of the 9 numbers. The result will be palindromic magic square.

181	111	161
131	151	171
141	191	121

The magic sum of this square is 453. We can continue this process further.

Instead of the 3-digit number we can convert each into a 4-digit number. Let us repeat the number in the ten's place.

1881	1111	1661
1331	1551	1771
1441	1991	1221

The magic sum of this square is 4653. Let us take it one step further and see what happens. Insert 1 in the hundred's place.

18181	11111	16161
13131	15151	17171
14141	19191	12121

The magic sum of this square is 45453.

Look at the following squares which are numbered from A to D.

A

444	999	222
333	555	777
888	111	666

magic sum = 1665

Take one digit of the above square (middle of each number) and include first and last arbitrary digits.

B

646	696	626
636	656	676
686	616	666

magic sum = 1968

Take 3-digits of A and insert arbitrary second and forth digits.

C

43434	93939	23232
33333	53535	73737
83838	13131	63636

magic sum = 160605

Take 1st three digits of C.

D

434	939	232
333	535	737
838	131	636

magic sum = 1605

How to Produce Palindromes?

A normally quick way to produce a palindromic number is to pick a positive integer of 2 or more digits, reverse the digits and add to the original, then repeat this process with the new number and so on.

e.g. let us take 3462.

3462	6105	11121
2643	5016	12111
6105	11121	23232

Let us consider two more examples where we can produce a palindrome in one step.

2010	2013
0102	3012
2112	5115

Does the series formed by adding a number to its reverse always end in a palindrome? It used to be thought so. However, this conjecture has been proven false for base 2, 4, 8 and other powers of 2 seems to be false for base 10 as well.

Palindromic numbers can be prime, composite, odd, even, square, cube, and so on. Squares are the squares of 1,11,111,1111 etc which are 1, 121, 12321 and 1234321.The number 26 is the smallest non-palindrome number whose square happens to be palindrome number 676.

Some palindromic cubes derived from the cubes of 7, 11, 101, 111, 1001, 10101 etc. are themselves palindrome.

A special palindrome prime is the following 13-digit number.

1888081808881

This number reads the same upside down or when viewed in the mirror.

Look at the following 3X3 magic square. All the numbers are palindromes. If you look at this square upside down, it is still magic. Hence 110 is written as 0110 and 220 is written as 0220 for symmetry in the result.

Look at the following 3X3 magic squares using palindromes.

1221	0000	2112
2002	1111	0220
0110	2222	1001

282	737	646
919	555	191
464	373	828

The magic sum of this square is 1665. Note that opposite cells add up to 1110. Also note that 1665 = 666+333+666!

282+828 = 1110

464+646 = 1110

Also note that 1110 = 666+444

212	767	656
989	545	101
434	323	878

The magic sum of this square is 1635. Note that 1635 = 666+303+666 !

Look at the following 3X3 palindrome magic square with centre cell as 545 similar to the one given above.

636	181	818
727	545	363
272	909	454

The magic sum of this square is 1635.

The above square can be transformed to another palindromic magic square by adding 2 zeros to the appropriate places in between.

60306	10801	80108
70207	50405	30603
20702	90009	40504

The magic sum of this square is 151215

The following of the 3-digit magic square has all the palindromic numbers. It is known as Beastly magic square.

212	121	333
343	222	101
111	323	232

The magic sum of this square is 666. This is known as the number of the Beast. Now you know why it is called as Beastly magic square.

Try to solve the following problem.

In the following magic square all the digits 1 to 9 are to be used only once.

a	b	c
d	e	f
g	h	i

Across: abc, def, ghi   Down: adg,beh,cfi

Find all the 3-digit number formed so that

1)adg +beh+cfi is a palindrome and abc +def +ghi is a square of the palindrome.

2) adg+beh+cfi is a palindrome and abc +def +ghi is twice the palindrome.

3) adg +beh+ ghi is a palindrome and abc+ def +ghi is a palindrome.

Look at the following 3 magic squares. They involve use of 27 distinct palindromes.

272	353	161		505	343	727		777	696	888
151	262	373		747	525	303		898	787	676
363	171	252		323	707	545		686	878	797

The magic sums of the 3 squares are 786, 1575 and 2361. The magic sum of the last square 2361=3 (magic sum of the first square) +3

(786x3)+3=2361

There is a peculiarity with this number 2361. This number in the decimal system when written in hexadecimal system, gives another 3-digit palindrome.

$$2361 \quad = \quad 939$$
$$\{10\} \qquad \{16\}$$

Magic squares composed of palprimes all with 9-digits.

147343741	125939521	139626931
129919921	137636731	145353541
135646531	149333941	127929721

162484261	125474521	145494541
127494721	144484441	161474161
143474341	163494361	126484621

177565771	131555131	157878751
135979531	155666551	175353571
153454351	179777971	133767331

11-Digit Palindromic Primes

10797779701	14336063341	12568886521
14338283341	15266746521	10796669701
12566366521	10798889701	14337173341

The above square contains 3 triplets 10796669701, 10797779701 and 10798889701. The common difference is 1110000.

The following 3 squares have same number in the central cell.

18264546281	14428082441	16670707661
14860606841	16454445461	18048284081
16238183261	18480808481	14644344641

17744444777	14261716241	17357175371
16067176061	16454445461	16841714861
15551715551	18647574681	15164446151

17762726771	14054145041	17546464511
16238183261	16454445461	16670707161
15362426351	18854745881	15146164151

13-Digit Palprimes

1026438346201	1000008000001	1013538353101
1000434240001	1013328233101	1026228226201
1013118113101	1026648466201	1000218120001

27-Digit Palindromes

Nine numbers are given below which are in arithmetic progression and all are 27-digit palprimes. Arrange these 9 numbers in the form of 3X3 magic square.

159056556140202041655650951

159056556141212141655650951

159056556142222241655650951

159056556143232341655650951

159056556144242441655650951

159056556145252541655650951

159056556146262641655650951

159056556147272741655650951

159056556148282841655650951

The common difference between these numbers is 1010100000000000 which is a 11- digit number.

92

Chapter 13

HETERO AND SEMIMAGIC SQUARES

Like magic squares, there are arrangements of numbers similar to magic squares that have interesting properties. These arrangements are called hetero squares and semimagic squares.

Heterosquares

Unlike magic squares, in hetero squares the numbers do not add up to the same value (magic constant), when you sum the rows, columns and diagonals. Rather, the sums of the rows, columns and diagonals are all different from each other.

We use the integers 1 to n^2 similar to what we did for magic squares of order 3. Hetero squares exist for $n \geq 3$. Depending upon whether n is odd or even, the method of construction differs. In a hetero square, the rows, columns and diagonals sum to different values.

Let us construct are 3X3 hetero square. Because n is odd, we fill the square in a spiral pattern.

- Clockwise

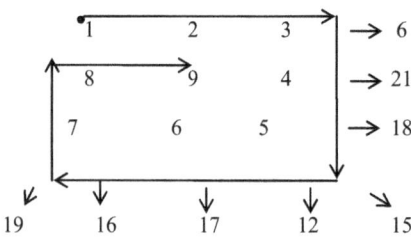

- Anticlockwise

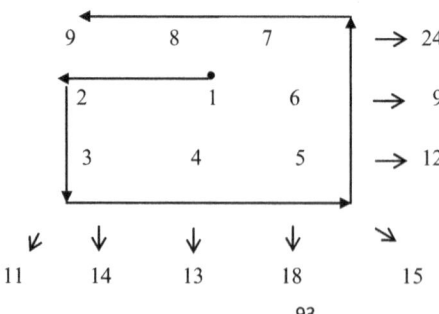

93

Hetero square can be considered as the black sheep of magic squares. Using the numbers 1 to 9, it has been shown that there are 3120 different basic (no rotations or reflections) hetero squares of order 3.

Prime Hetero Squares

Let us take the following prime numbers.

3, 5, 7, 11, 13, 17, 41, 47, 83

3, 5, 7, 11, 17, 23, 29, 31, 37

Two hetero squares can be formed with these prime numbers.

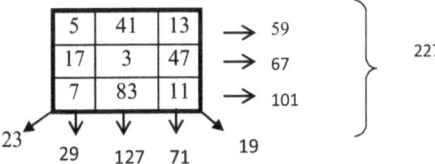

3 rows, 3 columns and 2 diagonals sum to prime numbers. Sum of all 9 cells = 227, a prime.

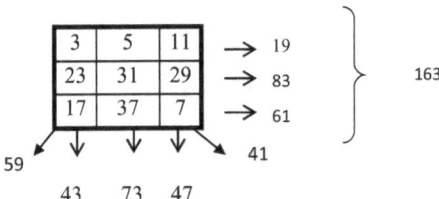

Sum of all 9 cells = 163, a prime.

Consider the 9 consecutive primes from 31 to 67. i.e 31, 37, 41, 43, 47, 53, 59, 61, 67. Look at the following 6 hetero squares.

31	37	41	→ 109
53	59	61	→ 173
67	43	47	→ 157
			439

31	43	53	→ 127
59	67	37	→ 163
61	47	41	→ 149
			439

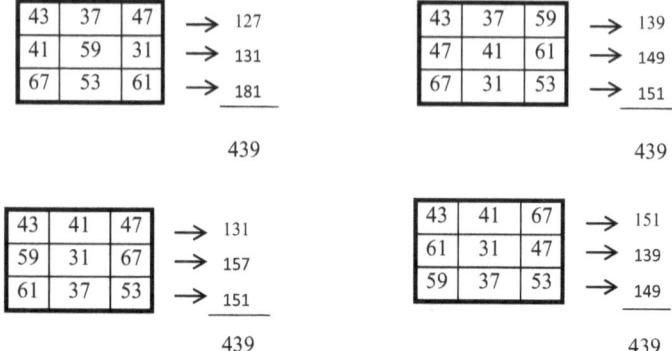

43	37	47
41	59	31
67	53	61

→ 127
→ 131
→ 181

439

43	37	59
47	41	61
67	31	53

→ 139
→ 149
→ 151

439

43	41	47
59	31	67
61	37	53

→ 131
→ 157
→ 151

439

43	41	67
61	31	47
59	37	53

→ 151
→ 139
→ 149

439

If you add all the 9 squares in the hetero square you will get the number 439 which is prime.

8 line sums are all primes in the range 109 to 173.

109,137,139,149,151,157,167,173 are eight consecutive primes.

Let us have some more fun with hetero squares. Look at the 3X3 hetero square given below.

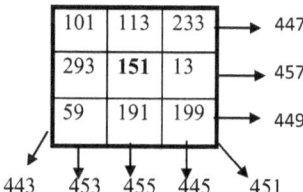

101	113	233
293	**151**	13
59	191	199

→ 447
→ 457
→ 449

443 453 455 445 451

The line sums range from 443 to 457. The centre number 151 is a pal prime (a palindrome prime number).Let us look at one more example of a 3X3 hetero square.

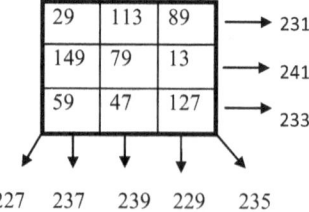

29	113	89
149	79	13
59	47	127

→ 231
→ 241
→ 233

227   237   239   229   235

The line sums in this hetero square range from 227 to 241.

Semimagic Squares

In a magic square, the rows, columns and diagonals sum to the same value. If one or both the diagonals do not sum to the same value, it is called as a semi magic square.

Using the numbers 1 to 9, we give below 8 semimagic squares.

<u>One diagonal fails to give the same sum</u>

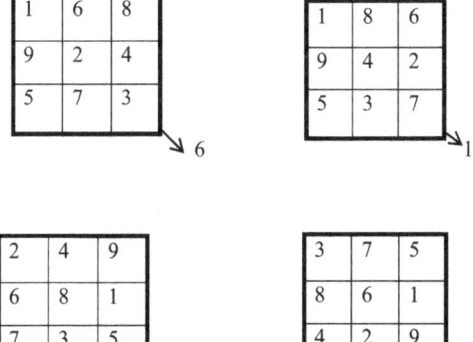

1	6	8
9	2	4
5	7	3

↘ 6

1	8	6
9	4	2
5	3	7

↘12

2	4	9
6	8	1
7	3	5

24 ↙

3	7	5
8	6	1
4	2	9

↘ 18

Two Diagonals fail to give the Same Sum

18 ↗

1	5	9
6	7	2
8	3	4

24 ↙   ↘ 12

1	5	9
8	3	4
6	7	2

↘ 6

18

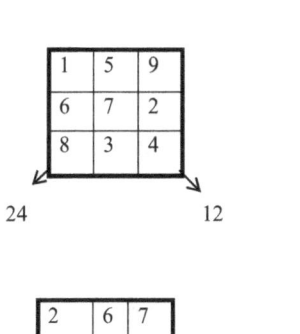

2	6	7
9	1	5
4	8	3

12 ↙   ↘ 6

3	4	8
5	9	1
7	2	6

24 ↙   ↘ 18

In real life situations, some problems relating to division of objects equal in numbers and value can be easily solved by constructing semimagic or magic square in accordance with the given conditions.

Matrix multiplication of 2 magic squares

If $A = \begin{pmatrix} 8 & 1 & 6 \\ 3 & 5 & 7 \\ 4 & 9 & 2 \end{pmatrix}$ and $B = \begin{pmatrix} 6 & 1 & 8 \\ 7 & 5 & 3 \\ 2 & 9 & 4 \end{pmatrix}$

Then,

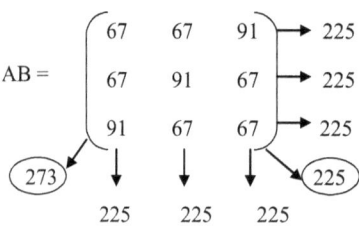

$AB = \begin{pmatrix} 67 & 67 & 91 \\ 67 & 91 & 67 \\ 91 & 67 & 67 \end{pmatrix}$

One diagonal fails. Hence AB is not a magic square, but of course it is a semi magic square.

Addition of 2 Semimagic Squares

When 2 semi magic squares are added, the result is a semi magic square.

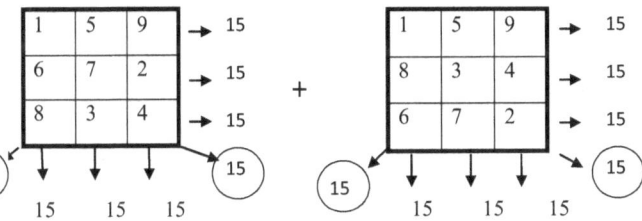

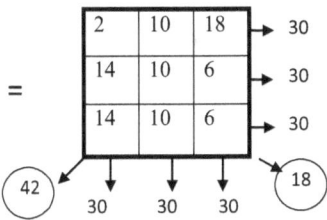

3X3 Multiplicative Semimagic Squares

See the following examples. Instead of addition, we do the multiplication and call the result as magic product.

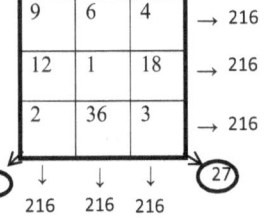

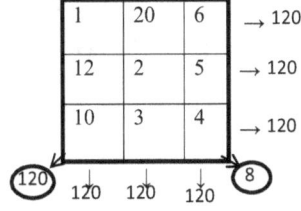

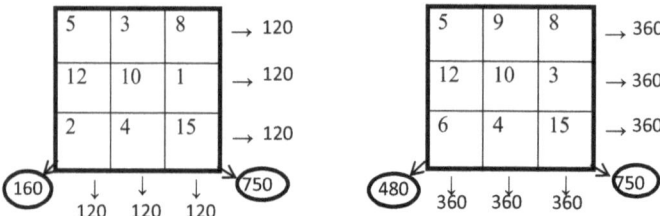

The smallest possible semimagic product of 3X3 square is 120. Can you prove this?

3X3 Semimagic Square of Squares

Look at the following semi magic square which consists of numbers in squares.

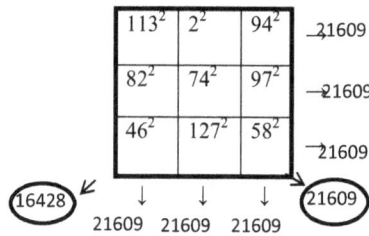

Only the diagonal from left top to bottom right is correct. The other diagonal, top right to bottom left gives a different total.

79^2	89^2	91^2
35^2	65^2	85^2
13^2	23^2	47^2

All sums going through the centre have the same value. The other 4 line sums do not match.

Semimagic Square of Cubes

Consider the numbers 1 to 9. Using the cubes of these integers can you construct a semi magic square? If not, can you prove that such a square cannot exist?

Such a construction is not possible. It can be seen in the following example.

The magic sum of the cubes is 675 and 9 needs to be one of the entries. But, the cube of 9 is 729, too big to be in a magic series summing to 675.

$1+8+27+64+125+216+343+512+729$

Semimagic Squares of Triangular Numbers

Triangular numbers come from the number of dots in a right angled triangle diagram like this one:

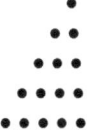

They are formed by adding up the series

$1+2+3+4+5.......$

To find out, say, the nth triangular number or T_n, use the formula $T_n = \frac{1}{2}(n (n+1))$

So, $T_5 = \frac{1}{2} (5(5+1))$ $= \frac{1}{2} (5 \times 6) = 15$

In the following semi magic square there are 9 distinct triangular numbers.

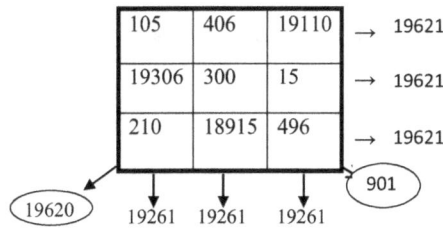

The magic sum of this square is 19621. One of the diagonals sums to 19620 = S-1. The other diagonal sums to 901. All the rows and columns sum to the same value.

Semimagic Squares of Pentagonal Numbers

Pentagonal numbers come from the 5-sided figure, the pentagon. Keep expanding by putting another pentagon on the outside. The pentagonal numbers are the total number of points in each stage of the diagram.

The first few pentagonal numbers are 1,5,12,22,35,51,70,92,117,145,176,210,247,.........

The nth pentagonal number

$P_n = 3n^2 - n/2$

The nth pentagonal number is one third of the $3n-1^{th}$ triangular number.

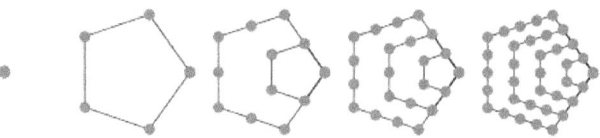

Look at the following semimagic square. It consists of distinct pentagonal numbers.

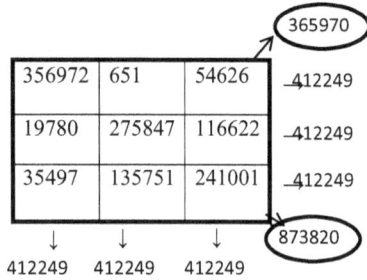

356972	651	54626	↗ 365970
356972	651	54626	⌐412249
19780	275847	116622	⌐412249
35497	135751	241001	⌐412249
↓	↓	↓	873820
412249	412249	412249	

The diagonals do not match.

Chapter 14
PRIME NUMBERS

Primes are a special section of whole numbers. Mathematicians have been asking questions about prime numbers for about 25 centuries and every answer seems to generate a flurryof new questions. Throughout the history of prime numbers all gifted mathematicians from Pythagoras and Euclid to Fermat, Gauss and 20^{th}century mathematicians including Ramanujan figure prominently.

A prime number (or a prime) is an integer greater than 1 that can be divided only by itself and 1.A natural number greater than 1 that is not a prime number is called a composite number.In the olden days 1 was considered as a prime number. But, it is no longer so. So, the first prime number and the only to be even is 2. It is followed by 3,5 and 7.See the table below which gives the prime numbers from 1 to 1000.

2, 3, 5, 7, 11, 13, 17, 19, 23, 29, 31, 37, 41, 43, 47, 53, 59, 61, 67,

71, 73, 79, 83, 89, 97, 101, 103, 107, 109, 113, 127, 131, 137, 139, 149,

151, 157, 163, 167, 173, 179, 181, 191, 193, 197, 199, 211, 223, 227, 229,

233, 239, 241, 251, 257, 263, 269, 271, 277, 281, 283, 293, 307, 311,

313, 317, 331, 337, 347, 349, 353, 359, 367, 373, 379, 383, 389, 397, 401,

409, 419, 421, 431, 433, 439, 443, 449, 457, 461, 463, 467, 479, 487,

491, 499, 503, 509, 521, 523, 541, 547, 557, 563, 569, 571, 577, 587, 593,

599, 601, 607, 613, 617, 619, 631, 641, 643, 647, 653, 659, 661, 673, 677,

683, 691, 701, 709, 719, 727, 733, 739, 743, 751, 757, 761, 769, 773,

787, 797, 809, 811, 821, 823, 827, 829, 839, 853, 857, 859, 863, 877, 881,

883, 887, 907, 911, 919, 929, 937, 941, 947, 953, 967, 971, 977, 983, 991, 997

Number 5 for example, is a prime as it is divisible by only 1 and 5, whereas 6 is composite because it has the divisors 2 and 3 in addition to 1 and 6.This division between prime and composite numbers turnout to be the corner stone of mathematics. There are an infinite number of prime numbers. Another way of saying is that the sequence 2,3,5,7,11,13,.....of prime numbers never ends.

Mersenne primes

Martin Mersenne (1588-1648) the seventh century monk, French philosopher and mathematician searched for the formula to generate prime numbers. Look at the formula.

$$2^n-1$$

Yes, when n = 2, 2^2-1 = 3 and 3 is a prime number. Likewise, 2^3-1 = 7 is a prime number. But the formula parts 2^4-1 = 15. 15 is not a prime number.

He thought this way, if I limit n to prime numbers only? If p is a prime, will 2^p-1 also be a prime? The answer is sometimes yes and sometimes no.

$$2^5-1 = 31; \quad 2^7-1 = 127; \quad 2^{11}-1 = 2047$$

But 2047 = 23x89 so $2^{11}-1$ is not a prime.

$2^{13}-1$ = 8191, $2^{17}-1$ = 131071 and $2^{19}-1$ = 524287 are all primes.

Fermat was a friend of Mersenne. He conjectured that the numbers 2^n+1 was always prime. It works from 2,4,8 and 16 giving prime numbers 5,17,257 and 65,537, but 100 years later, Euler showed that $2^{23}+1$ =4,294,967,297 was not a prime number because it is divisible by 641.

Primes once the exclusive domain of pure mathematics have recently found an unexpected ally in matters of computer security. Based on the difficulty of factorizing a product of 2 very large primes, public key cryptography was invented.

Recently in January 2018, the 50th Mersenne prime was discovered. Its value is calculated by raising 2 to 77, 232, 917th power and subtracting 1.

Palindromic prime

A palindromic prime is a prime number which is also a palindromic number. The first few palindromic primes are 2,3,5,7,11,101,131,151,181,191,313, 353, 373, 383, 727, 757, 797, 919, 929........ Except for 11, all palindromic primes have an odd number of digits.

Twin prime

A twin prime is a prime number that differs from another prime number by 2. Except the pair (2,3) this is the smallest possible difference between the two primes. Some examples of twin primes are given below.

(3,5), (5,7), (11,13), (17,19), (29,31) and (41,43).

Magic squares containing only prime numbers

The following four 3X3 magic squares contain only prime numbers.

461	23	269
59	251	443
233	479	41

1181	23	719
179	641	1103
563	1259	101

2621	23	1439
179	1361	2543
1283	2699	101

2381	1439	1973
1523	1931	2339
1889	2423	1481

3X3 magic square with central values less than 100

If we can construct the 3X3 magic square with centre value less than 100, then we can have solutions which are given below.

101	5	71
29	59	89
47	11	107

101	29	83
53	59	89
59	113	41

109	7	103
67	73	79
43	139	37

149	11	107
47	89	131
71	167	29

If we can consider prime under 1000 to be the centre value, then there are 474 possible solutions. If any prime number under 1,000,000 is considered then there are 1044538640 possible solutions. The last prime number under 1000000 is 999983. There are 33542 possible 3x3 prime number magic squares with value of 999983 in the centre square. One of them is given below.

1009,373	982,343	1,008,233
998,843	999,983	1,001,123
991,733	1,017,623	990,593

104

Emirps

13 has never been a popular number but recently it has found a new fame for being the first emirp. An emirp (prime spelled backward) is a prime number whose reversal is also a prime. This does not include palindromic prime like 11 or 101. The list starts with 13, 17, 31,37,71,73,79,97, 107,113,149,157,167,179,199,311,337,347,359.... and so on.

In Harry Nelson's spectacular magic square of 10-digit primes, only one is emirp. Can you find out from the following square?

1480028159	1480028153	1480028201
1480028213	1480028171	1480028129
1480028141	1480028189	1480028183

The following 3X3 magic square has all the entries as emirps.

17207	18731	16493
16763	174777	18191
18461	16223	17747

The magic constant of this square is 17477x3 =52431. The rows, diagonals and columns satisfy this requirement.

Six 4- digit primes are given below; 9721,9749,9769,9781,9787,9791. They are very far to each other on the number line. They all have emirpy partners.

Sexy primes

The Latin word for the number 6 is 'sex', as in sextuplets. You have a sexy prime pair when adjacent prime numbers differ by 6. It is thought, but not yet proved that there is an infinite number of them along the number line. Look at the following pairs.

23 and 29; 31 and 37; 53 and 59; 61 and 67; 73 and 79; 83 and 89.

Earlier we had mentioned that in Harry Nelson's square there is one emirp. In that same square there are 2 spectacular sexy prime pairs. Can you find them?

Chen Primes

A prime number p is called a Chen prime, if p+2 is either a prime or a product of two primes (also called a semi prime). This name was assigned in recognition of Chen's theorem that every sufficiently large even number can be written as the sum of a prime and a semiprime.

A semiprime is a natural number that is the product of two (not necessarily distinct) prime numbers. The following list shows semiprimes less than 100.

4,6,9,10,14,15,21,22,25,26,33,34,35,38,39,46,51,55,57,58,62,65,69,74,77,82,85,86,87,91,93,94,95.

By definition, semiprime numbers have no composite factors other than themselves. For example 26 is a semiprime and its only factors are 1, 2, 13 and 26.

Let us see two examples. 41 is a Chen prime since 41 +2 = 43 is also a prime. But 43 is not a Chen prime because 45 has more factors to be a semiprime. 47 is a Chen prime since 49, the square of a prime, is a semiprime.

Chen Jingrun proved that there are infinitely many Chen primes. If we look at say p<100, it would appear that there are more Chen primes than non-Chen primes. Green and Tao proved that there are infinitely many Chen primes in arithmetic progression.

The first few Chen primes are : 2,3,5,7,11,13,17,19,23,29,31,37,41,47,53,59,67,71,83,89, 101,.......

The first few Chen primes that are not the lower member of a pair of twin primes are: 2,7,13,19,23,31,37,47,53,67,83,89,109,113,127,.......

The first few non-Chen primes are : 43,61,73,79,97,103,151,163,173,193,223,229,241,.......

Using 9 Chen primes can we constant a3X3 magic square? Yes. This was achieved by Rudolf Ondrejka. It is given below.

17	89	71
113	59	5
47	29	101

The magic constant for this square is 177. It consists of 3 arithmetic progressions with a common difference of 12. The arithmetic progressions are:

5,17,29; 47,59,71; 89, 101,113.

Sophie Germain Primes

If both p and 2p+1 are prime then p is a Sophie Germain prime. It was named after the French mathematician Sophie Germain who used these in her investigations of Fermat's last theorem.

The first few such primes are 2, 3, 5, 11, 23, 29, 41, 53, 83, 89, 113 and 131. Take 29. It is a Sophie Germain prime because (2x29)+1=59. The number 59 is associated safe prime. Let us make this clear.

A safe prime is a prime number of the form 2p+1, where p is also a prime. Conversely, the prime p is Sophie Germain prime. The first few safe primes are

5, 7, 11, 23, 47, 59, 83, 107, 167, 179, 227, 263, 347, 359, 383, 467, 479, 503, 563, 587, 719, 839, 863, 887, 983, 1019, 1187, 1283, 1307, 1319, 1367, 1439, 1487, 1523, 1619, 1823, 1907,……..

With the exception of 7, a safe prime q is of the form 6k-1. With the exception of 5, a safe prime q is of the form 4k-1. Consequently a safe prime q>1 also must be of the form 12k-1.

These primes are called "safe" because of their relationship to strong primes. A prime number q is a *strong* prime if q+1 and q-1 both have large prime factors.

Sophie Germain primes and safe primes have applications in public key cryptography and primality testing.

3X3 Magic Squares using Sophie Germain Primes

Let us look at 2 examples

106121	179	55733
3623	54011	104399
522289	107843	1901

Magic Sum = 162033

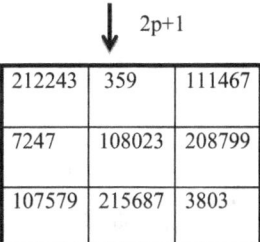

$2p+1$

212243	359	111467
7247	108023	208799
107579	215687	3803

Magic Sum = 324069

1481	1889	2063
2393	1811	1229
1559	1733	2141

Magic Sum = 5433

$2p+1$

2963	3779	4127
4787	3623	2459
3119	3467	4283

Magic Sum = 10869

We have just seen p and 2p+1. Now let us see p and 2p-1

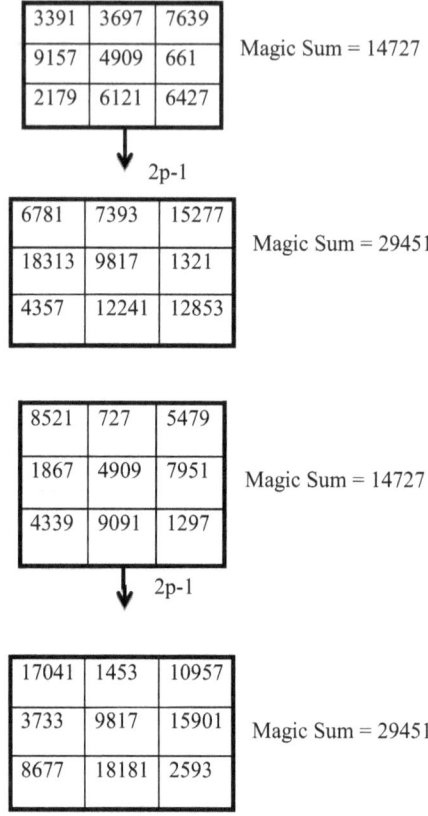

3391	3697	7639
9157	4909	661
2179	6121	6427

Magic Sum = 14727

2p-1

6781	7393	15277
18313	9817	1321
4357	12241	12853

Magic Sum = 29451

8521	727	5479
1867	4909	7951
4339	9091	1297

Magic Sum = 14727

2p-1

17041	1453	10957
3733	9817	15901
8677	18181	2593

Magic Sum = 29451

Can you construct 3X3 magic square using p and 4p+3 and p and 4p-3 where p is Sophie Germain prime?

Chapter 15

MAGIC NUMBER 1089

1089 is a square number $(33)^2$. It is widely used in magic tricks because it can be "produced" from any three digit number.

1. Take any 3-digit number in decreasing order where first and last digits differ by at least 2 or more
2. Reverse the number. Subtract the smaller from the larger.
3. Add the result to the number produced by reversing the digits.

Let us work out one example.

Take the 3-digit number 753; 753 - 357= 396; 396 + 693 = 1089

1089 and 3 X3 Magic Square

Let us consider the 3X3 magic square

4	9	2
3	5	7
8	1	6

Multiplying by 1089, we get

4356	9801	2178
3267	5445	7623
8712	1089	6534

This is a magic square whose magic sum is 16335.

From this magic square we can make several other magic squares.

Use the last 3 digits

356	801	178
267	445	623
712	089	534

The magic sum is 1335.

Use the last 2 digits

56	1	78
67	45	23
12	89	34

The magic sum is 135.

Write the first 3 digits

435	980	217
326	544	762
871	108	653

The magic sum is 1632.

Write the first 2 digits

43	98	21
32	54	76
87	10	65

The magic sum is 162.

Chapter 16

FIBONACCI SEQUENCE

Most of the number sequences are famous because they are very simple. Some of them exhibit very interesting properties. You already know about sequence of odd and even numbers. But, there is one number sequence that is more famous than any other and that is the one we will discuss in this chapter. It is called Fibonacci sequence after the mathematician who invented it.

Fibonacci was the most famous medieval mathematician. He was given a problem to solve in 1202. If two newly born rabbits, one male and one female are allowed to breed in ideal conditions, how many rabbits will be there at the end of 12 months? We have to assume that none of the rabbits die, they never escape from the field and always produce two babies in one month, one boy bunny and one girl bunny and that sibling mating is acceptable.

Fibonacci
1170-1250

Working on this problem, Fibonacci noticed that there was a pattern - each number was the sum of the previous two numbers. So, the sequence continues:

1,1,2,3,5,8,13,21,34,55,89,144,233,377,610 …. and so on. The 12th number along the line is 144 and so after one year there will be 144 rabbits in the field.

In this Fibonacci sequence, the term will be named as given below.

1st term	2nd term	3rd term	4th term	5th term
1	1	2	3	5

The general formula for the Fibonacci sequence is

$$F_{(n)} = F_{(n-1)} + F_{(n-2)}$$

Can you work out the 20th term for this Fibonacci sequence?

In the earlier example, we have started the sequence with 1. The sequence does not have to start with 1. It can start with any number as long as the principle employed is the same. See the examples given below.

1,3,4,7,11,18,29,47,76,....

2,3,5,8,13,21,34,55,89,144,233,377, ...

There is one very interesting fact about Fibonacci's numbers. When you divide one of the numbers from the sequence with the preceding number, you will get approximately the same number i.e. 1.618 which is known as the Golden ratio or Divine Ratio.

This number turns up in all kind of places and in all kinds of ways. Its value is

$$\Phi = (\tfrac{1}{2}) (\sqrt{5} +1)$$

It is an irrational number. This means that you cannot find its precise value on the calculator or in any other way! The Greek letter Phi is used to represent this. Its value is 1.618033 (rounded off to 6 decimal places).

The larger the Fibonacci numbers the closer this ratio of the last 2 terms approach the golden ratio.

$13/8 = 1.625; 55/34 = 1.61764...., 4181/2584 = 1.61803405 ...$

If you use the Golden ratio and the Fibonacci number sequence to draw geometrical figures you may end up constructing various figures. Phi can be found in pyramids, pentagram, distribution of leaves in the plants, relation between the length of our arm (fingers until shoulder) and the length until elbow.

There are no numbers in all of mathematics as all-pervading as the fabulous Fibonacci numbers. They pop up every now and then in nature, geometry, algebra, number theory, permutations and combinations and many other branches of mathematics.

Now let us turn our attention to magic squares. Can you form a magic square using Fibonacci numbers? There is apparently no 3x3 magic square with distinct Fibonacci numbers.

Recall the Fibonacci sequence. 1,1,2,3,5,8,13,21,34,..... Take the 3X3 magic square

6	1	8
7	5	3
2	9	4

You can use any of the 8 versions of this 3X3 square. Replace each number by its corresponding Fibonacci number.

F_6	F_1	F_8
F_7	F_5	F_3
F_2	F_9	F_4

$\longrightarrow$

8	1	21
13	5	2
1	34	3

Consider the products of the row numbers and products of the column numbers.

The sum of the row products and the sum of the column products are equal.
Now, let us take another Fibonacci series starting with 3.
3, 5, 8, 13, 21, 34, 5, 89, 144, ...

Let us take the fully 3X3 magic matrix which we call as A.

$$A = \begin{pmatrix} 4 & 9 & 2 \\ 3 & 5 & 7 \\ 8 & 1 & 6 \end{pmatrix}$$

Replace the entries in this magic square with the corresponding Fibonacci numbers. We end up with the matrix B

$$B \quad \begin{bmatrix} 13 & 144 & 5 \\ 8 & 21 & 55 \\ 89 & 3 & 34 \end{bmatrix}$$

It is like a multiplicative square using Fibonacci numbers. It is very interesting. Sum of the products of the 3 numbers in each of the 3 rows of B is equal to the sum of the products of the 3 numbers in each of the 3 columns of B.

Rows –   9360 + 9240 + 9078 = 27,678

Columns– 9256 + 9072 + 9350 = 27,678

This is not a true magic square. But there is some magic in it.

The matrix A is non-singular. This means that the determinant of the matrix A $\neq$ 0. Its inverse A^{-1} is fully magic. The Fibonacci matrix B is also non-singular, but its inverse B^{-1} ,however does not seem to have any magical properties.

8	1	21	168
13	5	2	130
1	34	3	102
140	170	126	400

$$\text{Matrix A} = \begin{pmatrix} 4 & 9 & 2 \\ 3 & 5 & 7 \\ 8 & 1 & 6 \end{pmatrix}$$

Determinant of A= 360

$$\text{Matrix B} = \begin{pmatrix} 13 & 144 & 5 \\ 8 & 21 & 55 \\ 89 & 3 & 34 \end{pmatrix}$$

Determinant of B = 663624

Let us look at the inverse of matrix A and B

$$A^{-1} = 1/360 \begin{pmatrix} 23 & -52 & 53 \\ 38 & 8 & -22 \\ -37 & 68 & -7 \end{pmatrix}$$

This is perfectly magic. The magic sum is 24.

$$B^{-1} = 1/221208 \begin{pmatrix} 183 & -1627 & 2605 \\ 1541 & -1 & -225 \\ -615 & 4259 & -293 \end{pmatrix}$$

This is not magic.

Let us look at some other interesting points about the Fibonacci sequence.

1. $1+1+2+3+5+8+13+21 = 55-1$ and
   $1+1+2+3+5+8+13+21 +34+55+89+144+233 = 610-1$

2. Twice any term added to the term before equals the next but one term.

   e.g $(2\times89) + 55 = 233$ and

   $(2\times233) + 144 = 610$

Fibonacci and Pythagoras were separated by about 17 centuries. You may be surprised to know that a little bit of mathematics brings them together.

Fibonacci sequence is

1,1,2,3,5,8,13,21,34,55,89,144,233,377,610,987,1597,2584, ...

Choose any 4 numbers block, say8, 13, 21, 34

Multiply the outer ones : 8 x 34 = 272

Multiply the inner ones : 13 x 21 = 273

Double your last answer : 273 x2 = 546

Let 272 and 546 be the shorter sides of a right angled triangle. Now calculate the length of the hypotenuse by applying the Pythagoras theorem.

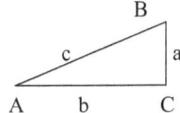

$c^2 = a^2 {}^+ b^2 = 272^2 + 546^2$

$$= 73984 + 29816$$

$$= 372100$$

$$C = \sqrt{372100}$$

$$= 610$$

Look at the sequence once more. You will find that 610 is further along the sequence.

The area of a triangle is ½ base x height

$$= (1/2 \times 272) + 546$$

$$= 74,256 \text{ square units}$$

Now multiply the 4 Fibonacci numbers you had chosen.

8 x 3 x 21 x 34 = 74,256

Is it not magic?

QUIZ

English Vocabulary – 1

Three quizzes are given. In each of them select the best definition for each of the words below. Put the number which corresponds to the letters of the vocabulary words in the magic square box. If the totals of the numbers are the same both across and down, you have found the magic number.

A) Elusive	1) often without awareness of some potential danger; self-satisfied
B) Complacent	2) not respectful; critical of what is generally accepted or respected
C) Amorphous	3) lacking in spirit or interest, listless, indifferent
D) Languid	4) a person who spreads alarming rumours
E) Tactile	5) perceptible to the touch
F) Supercilious	6) without shape or definite form
G) Scaremonger	7) having or showing arrogant superiority to and disdain of these one views as unworthy
H) Vacuous	8) hard to express or define; cleverly or skillfully evasive
I) Irreverant	9) lacking in ideas or intelligence

A	B	C
D	E	F
G	H	I

English Vocabulary – 2

A) Contraindicated	1) sleight of hand
B) Punctiliousness	2) thought experiment
C) gedanken experiment	3) employing big words
D) Prestidigitization	4) combination of visual images, usually different from sources
E) Disambiguation	5) clearing up confusion
F) Anthropomorphic	6) inadvisable
G) Insurrectionist	7) strict attention to detail
H) Sesquipedalian	8) rebel
I) Montage	9) having human characteristics

A	B	C
D	E	F
G	H	I

English Vocabulary –3

A) Symbiotic relationship when both organisms gain	1) carnivore
B) Something that eats a producer or another consumer to get energy	2) mutualism
C) An organisms that eats both plants and meat	3) parasitism
D) An organism that eats only plants	4) omnivore
E) A plant	5) producer
F) Symbiotic relationship when one organism gains and the other is hunt	6) sun
G) Most important link in the food chain	7) herbivore
H) An organism that eats only meat	8) commensalism
I) Symbiotic relationship when one organism gains and other is not bothered	9) consumer

A	B	C
D	E	F
G	H	I

Maths– 1

Solve each of the following equations. Write the solution to the first problem in square 1, the solution to the second problem in square 2 and so on. When you complete all the 9 problems correctly, you will have a magic square. Add up the numbers in each row, column and diagonal to check your answers.

1) $2x-1 = 69$	
2) $3x-7 = x+7$	
3) $-2x+18 = -x-9$	
4) $x-3/4 = x/5$	
5) $2x-1 = 45$	
6) $3x+5 = 98$	
7) $7+2x = 45$	
8) $4x-3 = 2x+75$	
9) $(x-1)/2 + (3x+2)/5 = 12$	

1	2	3
4	5	6
7	8	9

Maths– 2

Solve the following. The answers are given in the second column. Choose the correct answer and write the numeral appearing before the answer in the 3X3 square. If the answers are correct, you will get a magic square.

A) If a number is a multiple of 3, its digits add up to a multiple of ____	1) 10
B) If 6 is 24% of a number, what is 40% of the same number?	2) 5
C) What is the next number in the following pattern? 1, 1/2, ¼, 1/8,…..	3) 6
D) How long (in years) will you have to wait for your Rs. 2,500/- invested at 6% earn Rs. 600/- in simple interest?	4) 1.09
E) A product is on sale for Rs. 1600/-, which is a 20% discount off the regular price. What is the regular price?	5) 2000
F) Find 0.12 /12	6) 1/16
G) If Divya can do a job in 2 days, that Adit can do in 4 days and Swara can do in 6 days, how long (in days) would the job take if all of them work together to complete it?	7) 0.01
H) Swara needs to make a cake and cookies. The cake requires 3/8 cup of sugar and the cookies require 3/5 cup of sugar. Swara has 15/16 cup of sugar. Does she have enough sugar or how much more does she need?	8) 3
I) In a quiz there are 20 questions. Correct answer gives 5 points. Wrong answer gives – 2. A student gets 61 points. How many questions did he omit?	9) 3/80

A	B	C
D	E	F
G	H	I

Maths– 3

Solve the following. The answers are given in a 3X3 square. Write the number of the question pertaining to that answer. If the answers are correct, you will get a semimagic square. Do not bother about the diagonals. Add only the rows and columns. What is the magic sum you got?

1) The equation $x^2+6x+2=0$ cannot be solved by _____ the left side of the equation. A more powerful way to solve is needed.
2) $(x-2)(x+1)=0$ is equivalent to _____
3) An equation in the form of $ax+b=0$ is a _____
4) The fact that, if a product of 2 numbers is zero, then one or both the factors must be zero, allows us to solve any quadratic equation that has zero on one side provided we can ___ the other side.
5) $(x+3)^2=0$ is equivalent to _____
6) The roots of quadratic equation are _____ only if the discriminant is positive.
7)$ax^2+bx+c=0$ is the general form for a _____
8)$x = \dfrac{-b \pm \sqrt{b^2-4ac}}{2a}$
9) The expression b^2-4ac that appears under the square root sign in question 8 is called the _____

quadratic equation	linear equation	$x^2+6x+9=0$
$x^2-x-2=0$	factor	discriminant
real	quadratic formula	factoring

QUESTIONS

1) Complete the following magic square

5	8	-
-	12	-
-	-	-

The challenge is to find all the different 3X3 magic squares which contain 5, 8 and 12

2) Find a magic square whose first row is 0 8 4

3) Find a magic square whose first row is 1 8 3

4) Look at the following arithmetic sequences

1, 4, 7, 10, 13, 16, ……..

3, 8, 13, 23, 28, 33, …….

In the first sequence the common difference is 3 and in the next it is 5. In the following sequence can you tell what is the common difference?

25, 23, 21, 19, 17, 15,……………

5) Find the value of x in the magic square

34	-	-
-	x	-
-	-	56

6) For a magic square with the magic sum S, when a, b and c are placed as shown in the figure, prove that $C = \frac{S}{3} = \frac{a+b}{2}$

a	-	-
-	c	-
-	-	b

123

7) Triangular numbers are written as a sequence

1, 3, 6, 10, 15,…..

What are the next three in the sequence? What is the 12th term in the sequence?

8) Place the numbers 1 to 9 into a 3X3 grid such that the 4 sides and the 2 diagonals each have the same sum greater than 15. What is the number in the centre of the grid? Disregard rotations and reflections. Find all partially magic squares which use the numbers 1 to 9 and in which at least 6 lines of 3 numbers have same sum other than 15.

9) What number should be there in the centre?

1)

9	-	17
19	?	3
-	15	-

2)

5	-	13
-	?	-
9	7	-

10) Can you explain why 5 occupies the centre box in a 3X3 magic square consisting of numbers from 1 to 9?

11) Suppose you are given a 3X3 magic square M, with a magic sum=15

a) If you add 4 to each entry of M is the new square necessarily a magic square? If so what is the magic sum? If not, explain why not?

b) if you multiply each entry of M by 7 is the new square a magic square? If so what is the magic sum?

12) Write a computer program for playing with computer a 3X3 magic square. There is only one fundamental solution. However, if you include rotations and reflections there will be 8 solutions. Because the computer does not play a particularly creative game, all 8 solutions cannot be obtained. Can you modify your program to play a more interesting game which permits all 8 solutions.

13) Design a java program to generate a magic square and a method to print it. In addition, it should print the magic number. Your program should work in a loop, for more than just one magic square.

14) In a 3X3 magic square with numbers from 1 to 9, number 1 cannot appear as a corner element. Why?

15) You are given the following 8 numbers. 1, 7, 13, 31, 37, 43, 61 and 73. Find out the missing number and form the magic square.

16) A 3X3 multiplicative magic square has the property that the magic product K is the cube of the central square. The smallest magic product is 216. Can you prove these two statements?

17) You construct a 3X3 magic square using integers from 1 to 9.

Answer the following questions.

a) 5 occupies the centre square. Why?

b) You will notice that all the corners are occupied by even numbers. The odd numbers never appear in the corner of this magic square. Why?

c) The sum of the squares in the first row is equal to the sum of the squares in the third row. Similarly, the sum of the squares in the first column is equal to the sum of the squares in the last column. Can you explain why this happens?

d) The two numbers that are opposite each other across the centre number add up to the same number. Why?

e) Determine with proof, the central element of a 3X3 magic square made using the numbers 1, 2, 3,.......8, 9.

f) Prove that the magic sum of the square is 3 times the central number.

g) If you are asked to construct a 3X3 magic square using numbers 1 to 10 except 7, can you do it?

18) In a prison there are 9 cells. All prisoners can communicate with one another by doorways. The 8 prisoners have their numbers on their backs. Any of them is allowed to exercise himself in whichever cell may happen to be vacant, subject to the rule that at no time shall 2 prisoners be in the same cell.

One day they were given special comforts if without breaking that rule, they could place themselves that their numbers should form a magic square.

Now, prisoner number 7 happened to know a good deal about magic square; so he worked out a scheme and naturally selected the method that was most expeditious-that is one involving the fewest possible moves from cell to cell. But one man who was obstinate refused to move out of his cell or take part in the proceedings. But, number 7 was quite equal to the emergency and found that he could still do what was required in the fewest possible moves without troubling the brute to leave his cell.

The puzzle is to show how he did it and incidentally to discover which prisoner was so stupidly obstinate. Can you find the fellow?

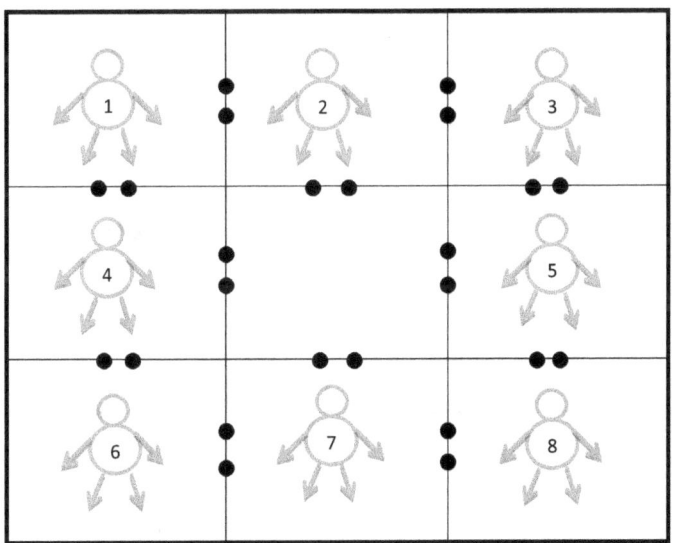

19) In the above problem, consider that of 9th prisoner was added to the vacant cell .They were offered complete liberty on the following strange conditions that their numbers formed a magic square without their movements causing any 2 of them ever to be in the same cell together except at the start one man was allowed to be placed on the shoulders of another man and thus add their numbers together and move as one man. For example, number 8 might be placed on the shoulder of number 2 and then they would move about together as 10.

You should solve this puzzle in the fewest possible moves and see that the man who is burdened has the least possible amount of work to do.

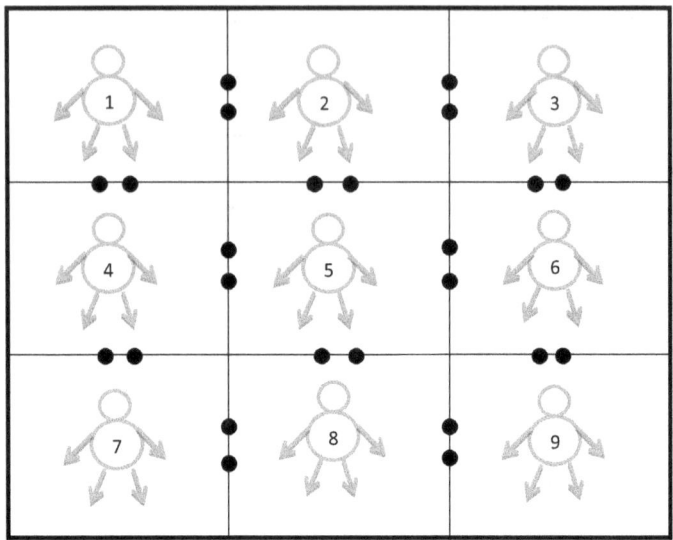

20) A fruit merchant had nine baskets . Every basket had plums .The number in every basket was different .When placed as shown in the illustration below ,they formed a magic square .So that if he took any three baskets in a line in the 8 possible directions there would always be the same number of plums.

The merchant told one of his men to distribute the contents of any basket he chose among some children , giving plums to every child so that each should receive an equal number.But , the man found it quite impossible , no matter which basket he selected and no matter how many children he included in the treat. Show,by giving the contents of the 9 baskets ,how this could come about. (note: There is a little trap in this puzzle).

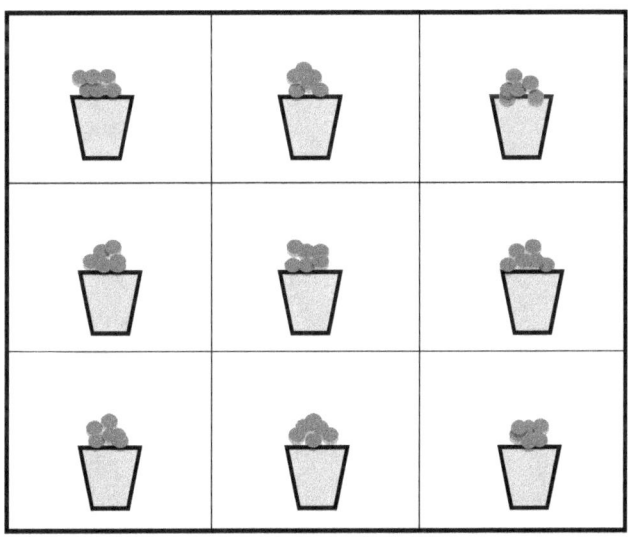

The above three problems are taken from the book 'Amusements in mathematics' by H.E.Dudeney. Try to solve all the three problems yourself. If you are unable to solve them, you may look at the solutions given in the book.

Construction of Odd Order Magic Squares

C++ Program

```
#include <stdio.h>
int main()
{
        int x;
l1:
        printf("Please Enter size of the Magic square to be created\n");
        scanf("%d",&x);
        if((x>2)&&((x%2)!=0))
        {
                int *cell= new int[x*x+1];
                for(int i=1;i<(x*x+1);i++)
                {
                        cell[i]=0;
                }
                int z,z1,z2,z3,z4,z5,z6;
                z=((x+1)/2)+(x-1);
                for(int y=1;y<(x*x+1);y++)
                {
                        z1=z-(int(z/x)*x);
                        if(z1==0)
                                z1=x;
```

```
        z2=((z-z1)/x)+1;

        if(z1==x)

                z3=1;

        else

                z3=z1+1;

        if(z2==1)

                z4=x;

        else

                z4=z2-1;

        z5=((z4-1)*x)+z3;

        if(z2==x)

                z6=1;

        else

                z6=0;

        if(cell[z5]!=0)

        {

                z3=z1;

                z4=(z2+1)-z6*x;

        }

        z5=((z4-1)*x)+z3;

        cell[z5]=y;

        z=z5;

}
printf("\nThe following magic square adds up to:%d\n\n",((x*x*x)+x)/2);

intcnt=0;

for(int b=1;b<=(x*x);b++)
```

```c
{
    printf("%d\t",cell[b]);
    cnt++;
    if(cnt==x)
    {
        printf("\n");
        cnt=0;
    }
}
}
else
{
    printf("\nThe size of square should be greater than 2 and it should be an odd number\n");
    printf("Please try another number\n");
    goto ll;
}

printf("\n");
printf("\n");
printf("\n");
}
```

Python Program

```
importnumpy as np
nb = input('Enter the order of magic square')
n=int(nb)
ms=[0]*(n*n)
z=((n+1)//2)+(n-1)
for i in range(1,n*n+1):
 z1=z-(int(z//n)*n)
if(z1==0):
  z1=n
 z2=(int(z-z1)//n)+1
if(z1==n):
  z3=1
else:
  z3=z1+1
if(z2==1):
  z4=n
else:
  z4=z2-1
 z5=((z4-1)*n)+z3
if(z2==n):
  z6=1
else:
```

```python
 z6=0

 z5=int(z5);

if(ms[z5-1]!=0):

 z3=z1

 z4=(z2+1)-(z6*n)

 z5=((z4-1)*n)+z3

ms[z5-1]=i

 z=z5

print ("The following magic square adds up to ", ((n*n*n)+n)//2)

print("\n")

for i in range (0,n):

for j in range(0,n):

print (int(ms[i*n+j]), end=' ')

print("\n")
```

www.ingramcontent.com/pod-product-compliance
Lightning Source LLC
Chambersburg PA
CBHW071316220526
45468CB00001B/392